Texts in Applied Mathematics 57

Hal Smith

An Introduction to Delay Differential Equations with Applications to the Life Sciences

 Springer

Hal Smith
School of Mathematical and Statistical Sciences
Arizona State University
Tempe, AZ
USA
halsmith@asu.edu

Series Editors

J.E. Marsden

Control and Dynamical Systems
107-81 California Institute of Technology
Pasadena, CA 91125
USA
marsden@cds.caltech.edu

L. Sirovich

Laboratory of Applied Mathematics
Department of Biomathematics
Mt. Sinai School of Medicine
Box 1012
New York, NY 10029-6574
USA
lawrence.sirovich@mssm.edu

S.S. Antman

Department of Mathematics
and
Institute for Physical Science
and Technology
University of Maryland
College Park, MD 20742-4015
USA
ssa@math.umd.edu

ISSN 0939-2475
ISBN 978-1-4614-2697-4   ISBN 978-1-4419-7646-8 (eBook)
DOI 10.1007/978-1-4419-7646-8
Springer New York Dordrecht Heidelberg London

Mathematics Subject Classification (2010): 34K05, 34K20, 92D25

Printed on acid-free paper

Springer is part of Springer Science+Business Media (www.springer.com)

Preface

This book is intended to be an introduction to delay differential equations for upper-level undergraduates or beginning graduate mathematics students who have a reasonable background in ordinary differential equations and who would like to get to the applications quickly. I used a preliminary version of this manuscript in teaching such a course at Arizona State University over the past two years. Existing texts on the subject by Diekmann et al. [26] and by Hale and Lunel [41], while excellent on the theory, are heavy on functional analytic background and light on applications. In my experience, most graduate students do not have the requisite background to read such texts profitably. A more applications oriented text by Kuang [48] is, unfortunately, out of print.

Both theory and applications of delay differential equations require a bit more mathematical maturity than its ordinary differential equations counterparts. Primarily, this is because the theory of complex variables plays such a large role in analyzing the characteristic equations that arise on linearizing around equilibria. Ideal prerequisites for this book include a second course in ordinary differential equations such as in the text [78, 10], some familiarity with complex variables, and some elementary analysis. Results from the calculus of several variables are routinely used, especially, the implicit function theorem.

This book focuses on the key tools necessary to understand the applications literature involving delay equations and to construct and analyze mathematical models involving delay differential equations. It begins with a survey of mathematical models involving delay equations. These are primarily from the biological literature, in keeping with my own prejudices, and due to the relative frequency of delay models in that literature relative to others. This is followed by a "warm-up" chapter on the simplest possible delay equation $u'(t) = -\alpha u(t - r)$. This simple example illustrates many of the complexities that arise with delays and has the advantage that results may be easily and explicitly worked out. Its main message is that delays naturally induce oscillations. Standard existence and uniqueness results are taken up in Chapter 3. The method of steps is introduced and exploited for discrete delay equations. For the reader interested mainly in applications, this may suffice. A more general approach follows but no fixed point theorems are used: the method of successive approximations works fine. A key notation is introduced here, one that takes a bit of getting used to, namely the state variable x_t which appears throughout the remainder of the book. In addition to continuous dependence of solutions on initial data, continuation of solutions, positivity, and comparison of solutions are also discussed because many applications come from biology where positivity restrictions are inherent to the models. Linear equations are taken up next with the primary aim being stability. In applications, linear delay equations arise through linearization of a nonlinear equation about an equilibria so the focus is on linear stability analysis and the characteristic equation the roots for which determine stability. Proof of the validity of linearized stability would require too much additional mathematics and therefore it is not given.

The following chapter is an introduction to abstract dynamical systems theory, using ordinary differential equations, discrete-time difference equations, and now delay differential equations as examples. It is shown that a delay differential equation induces a semidynamical system on the space of continuous functions on the delay interval. The focus then turns to omega limit sets, the usual results familiar from ODEs continue to hold but with some nuances due to the infinite-dimensional state space. Applications to the delayed logistic equation and the delayed chemostat model are treated. The LaSalle invariance principle is established and an application is given. Next, the Hopf bifurcation theorem, critical for applications, is treated. A simple canonical example is considered where the bifurcation can be explicitly computed. Following this, the Hopf bifurcation theorem is stated without proof. It is applied to the standard delayed negative feedback system $x'(t) = -f(x(t-1))$ where $xf(x) > 0$. In this case, a formal expansion for the periodic solution in terms of a small parameter (this is fully justified in an appendix) is given. Applications to various second-order delay equations are then considered, one of which is stabilizing the up position of a damped pendulum with delayed feedback; another is a model of a gene regulatory network. Finally, the beautiful Poincaré–Bendixson theory for monotone cyclic feedback systems, obtained recently by Mallet-Paret and Sell, is stated.

The following brief chapter is an introduction to equations with infinite delay and to the linear chain trick by which certain special kinds of infinite delays can lead to ordinary differential equations. These arise often in the modeling literature so an example is discussed in some detail. The final chapter focuses on a model of virus predation on a bacterial host in the setting of a chemostat where the bacteria subsist on a supplied nutrient. The delay corresponds to the latent period following virus infection during which new virus particles are manufactured within the cell. Most of the theoretical results of previous chapters are used in the analysis of this system of delay equations.

Two brief appendices should help those readers needing additional background on complex variables and analytic functions including the very useful Rouché's theorem, and implicit function theorems. The Ascoli–Arzela theorem is stated and discussed and the useful fluctuation method is described. A second appendix is devoted to a rigorous proof of Hopf bifurcation for the delayed negative feedback systems.

The impatient reader could skim the applications in Chapter 1, jump over Chapter 2, and start with Chapter 3. A note on notation: we use $\mathbb{R}$ for the set of real numbers, $\mathbb{C}$ for the set of complex numbers, and f' denotes the derivative of a function f.

I would like to acknowledge the influence of Yang Kuang, a collaborator on much of the author's own work in delay differential equations, on this work and to thank him for providing several figures used in the book. Several students, colleagues, and anonymous reviewers read portions of the manuscript and provided valuable feedback. Among these, the author would like to thank Patrick de Leenheer, Thanate Dhirasakdanon, Zhun Han, and Harlan Stech. Most of what I know about delay differential equations, I learned from Jack Hale, a giant in the field.

Finally, I have been supported by the NSF during the time this book took shape, recently by award DMS 0918440.

Tempe, Arizona   *Hal Smith*
                                                                                                                  July, 2010

Contents

Chapter 1
Introduction

Abstract Various delay differential equations, primarily taken from the biological sciences literature, are presented, along with necessary background from the application area, in order to motivate our study of delay differential equations. These range from models in population biology, physiology, epidemiology, economics, and neural networks, to control of mechanical systems. Some of these are treated in detail later in the text; others can serve as potential student projects. The reader should feel free to pick and choose to which of these to devote attention. Following this, some terminology is introduced and the reader is pointed to various computer software designed specifically for delay differential equations.

1.1 Examples of Delay Differential Equations

The familiar logistic equation describing the growth of a single population is given by
$$N'(t) = N(t)[b - aN(t)]$$
It assumes that population density negatively affects the per capita growth rate according to $dN/Ndt = b - aN(t)$ due to environmental degradation. Hutchinson [46] pointed out that negative effects that high population densities have on the environment influence birth rates at later times due to developmental and maturation delays. This led him to propose the delayed logistic equation

$$N'(t) = N(t)[b - aN(t - r)] \tag{1.1}$$

where $a, b, r > 0$ and r is called the delay. Throughout this book, $f'(t)$ denotes the derivative of function $f(t)$. See May [58], Nisbet and Gurney [62], and Ruan [63] for a description of the model. Wright's famous conjecture [80], made in 1955, regarding the solutions of (1.1) remains open; it is discussed in Chapter 5. Arino, Wang, and Wolkowicz [4] propose an alternative model that has simpler dynamics.

H. Smith, *An Introduction to Delay Differential Equations with Applications to the Life Sciences*, 1
Texts in Applied Mathematics 57, DOI 10.1007/978-1-4419-7646-8_1,
© Springer Science+Business Media, LLC 2011

Perhaps it would be more realistic to assume that the density dependence is distributed over an interval in the past rather than concentrated at a single time instant. This would yield:

$$N'(t) = N(t)[b - a \int_0^\infty N(t-s)k(s)ds] \qquad (1.2)$$

where kernel k is normalized so that $\int_0^\infty k(s)ds = 1$.

A criticism of the delayed logistic equation is that the birth and death rates are not clearly distinguished. Nicholson's data on population fluctuations of the sheep blowfly *Lucillia cuprina* motivated the model now referred to as the Nicholson's blowfly equation:

$$N'(t) = bN(t-r)\exp(-N(t-r)/N_0) - \delta N(t) \qquad (1.3)$$

The blowfly lays eggs that hatch larvae which ultimately become sexually mature flies. The model accounts only for adult flies for which food is supplied at a constant rate. It is assumed that eggs take exactly r time units to develop into adults. The first term on the right side accounts for recruitment of new adults; it should be regarded as a product of a fecundity term and a probability of survival from egg to adult: $aN(t-r) \times c\exp(-N(t-r)/N_0)$. The probability of survival decreases with increasing population size due to intraspecific competition for food among the immature flies. The final term on the right accounts for death. See [62, 58] for an extensive discussion.

Volterra (1931) introduced the predator-prey system

$$N_1'(t) = N_1(t)[b_1 - a_{12}N_2(t)] \qquad (1.4)$$
$$N_2'(t) = N_2(t)[-b_2 + a_{21} \int_{-\infty}^t N_1(s)k(t-s)ds]$$

where the fraction $k(t-s)$ of prey fish eaten at time $t-s$ is assumed to be translated into predator fish biomass at time t. Volterra's work, including the derivation of this model, is easily accessible in [67]. See Cushing for more on equations of this type [24].

Mackey models the control of carbon dioxide levels in the blood by the equation

$$x'(t) = \lambda - \alpha x(t)V_m \frac{x(t-\tau)^n}{\theta^n + x(t-\tau)^n} \qquad (1.5)$$

where $x(t)$ is the partial pressure of carbon dioxide in the blood and τ is the time between oxygenation of blood in the lungs and stimulation of chemoreceptors in the brainstem. See Glass and Mackey [36] and Keener and Sneyd [47].

The Mackey–Glass equation for the density of certain blood cells is famous for its chaotic behavior. It is

$$x'(t) = -\alpha x(t) + \frac{\beta x(t-\tau)}{x(t-\tau)^n + A^n} \qquad (1.6)$$

where $\alpha, \beta, A, \tau > 0$ and τ is the delay between initiation of cellular production in the bone marrow and release of mature cells into the blood. See Glass and Mackey [36]. This same equation has been employed to model cortisol concentrations in the blood of human subjects by Dokoumetzidis et al. [54, 27]. In order to take account of the 24 circadian rhythm influencing cortisol secretion, they take $\beta = kA^n$, where

$$A = A_{max} \cos\left((t - f)\frac{2\pi}{1440}\right) + B.$$

Cortisol, often termed the "stress hormone", is secreted from the adrenal gland. It indirectly controls its own secretion in a complicated feedback cycle, the duration of which is the delay τ in the model.

More sophisticated models of the production of red blood cells by the stem cells in the bone marrow, termed erythropoeisis, have been developed by Bélair, Mackey, and Mahaffy [5]. See also [47]. The hormone erythropoiten, secreted by cells in the kidney, stimulates the production of precursor cells that will eventually mature into red blood cells. Therefore, a model should include this hormone as well as blood cell densities. Bélair et al. obtain the following system for erythropoiten E and mature blood cells M,

$$M'(t) = r[S(E(t - T_1)) - e^{-\gamma T_2} S(E(t - T_1 - T_2))] \qquad (1.7)$$
$$-\gamma M(t)$$
$$E'(t) = f(M(t)) - kE(t)$$

Function $S(E)$, depending on erythropoiten level E, is the recruitment rate of blood cell precursor cells. It is an increasing function of its argument that vanishes when $E = 0$; in simulations described in [5], it is taken to be linear. Precursor cells require time $T_1 \approx 5 - 9$ days to mature, hence the first delay appearing in the equation for mature cells represents the time for precursor cells to mature. $T_2 \approx 120$ days is the assumed maximum age of a mature cell. Parameter γ is the decay rate of mature cells, and r is a composite parameter. Roughly, mature cells increase due to recruitment of precursor cells produced at time $t - T_1$, which mature at time t, and are lost due to a background decay rate and abruptly when the mature cells reach age T_2. The hormone production rate $f(M)$ is a decreasing function of mature cell density, typically taken to be a Hill function $f(M) = a(a + bM^p)^{-1}$, and the hormone is degraded in the liver at rate k. These monotonicity assumptions allow us to see the effect of a sudden decrease in red blood cells, for example, due to blood donation, assuming the system was formerly in equilibrium. In response, erythropoiten production is increased leading to higher concentration levels that stimulate production of precursor cells. Following the delay time T_1, these cells mature into new mature cells causing a rebound in mature cell density. This, in turn, reduces erythropoiten production.

Actually, Bélair et al. [5] derive a much more general partial differential equation model of erythropoeisis that describes the densities of precursor cells and mature cells with respect to cell maturity and time, from which they obtain the system (1.7) under special assumptions. This paper is especially recommended for its careful derivation of the delay differential equations from first principles. Simulations of (1.7) show oscillatory return to equilibrium following blood donation; sustained oscillations can also occur under suitable conditions that may reflect a diseased state.

More recent work on the regulation of blood cell types involving delay differential equations may be found in [18, 21] and the references therein.

Brunovsky et al. [11] model the deviation $x(t)$ of the value of a foreign currency from an assumed constant baseline level that may not be precisely known by currency traders by the equations

$$x'(t) = a[x(t) - x(t-1)] - b|x(t)|x(t) \tag{1.8}$$

where $a, b > 0$. Traders look at the previous trajectory of the currency to estimate what it will do in the future. If it is currently higher than it was $(x(t) > x(t-1))$ then there is a tendency to predict a continual rise in its value and vice-versa if it is presently lower than it was in the past. This is the meaning of the first term on the right side. But there are bears lurking in the woods if its value gets too high (or low). They will feel that it must come down because it surely must be higher than its true value. This is the meaning of the last term. Brunovsky et al. [11] show that for $a < 1$ the $x = 0$ equilibrium is asymptotically stable whereas for $a > 1$ there is a nonzero periodic solution that does not arise as a Hopf bifurcation.

An $S \to I \to R$ epidemic model with fixed period of temporary immunity is given by

$$S'(t) = -\beta I(t)S(t) + \gamma I(t - \omega)$$
$$I'(t) = \beta I(t)S(t) - \gamma I(t) \tag{1.9}$$
$$R'(t) = \gamma I(t) - \gamma I(t - \omega)$$

S denotes susceptible individuals, I denotes infectives, and R recovereds. Note that an individual remains in the R class precisely ω units of time: $\beta, \gamma, \omega > 0$. See Brauer and Castillo-Chavez [9]. This delay system must be treated with care in order that R not become negative! One can also add birth and death to the model by adding $\mu - \mu S$ to the first equation and subtracting μI from second and μR from third equation where μ is the birth (death) rate.

Busenberg and Cooke [12] introduce the following periodic delay equation

$$y'(t) = b(t)y(t - T)[1 - y(t)] - cy(t) \tag{1.10}$$

for the proportion of infectious individuals with a communicable disease carried by a vector, such as a mosquito. Humans are infected by contact with an infected vector and susceptible vectors are infected by contact with an infected person, becoming able to infect a susceptible human after a delay T during which the infectious agent develops. Seasonal periodic incidence is captured by assuming that $0 \le b(t) = b(t + \omega)$; for example $b(t) = b(1 + a\sin(2\pi t/\omega))$, $0 < a < 1$. Parameter c is the recovery rate.

De Gaetano and Arino [25, 60] model the intravenous glucose tolerance test using

$$G'(t) = -aG(t) - bI(t)G(t) + c(t) \qquad (1.11)$$

$$I'(t) = -dI(t) + e \int_{t-\tau}^{t} G(s)\,ds$$

where G is glucose blood concentration and I is insulin blood plasma concentration. The integral term represents the pancreatic release of insulin which depends on the distribution of glucose over the past τ time units. $c > 0$ is a glucose source coming from liver secretion; both insulin and glucose are removed from the blood by cellular uptake. For more current work on the modeling of insulin see Li and Kuang [52].

A simple model of a single self-excited neuron with delayed excitation is given by

$$x'(t) = -\alpha x(t) + \tanh(x(t - \tau)) \qquad (1.12)$$

where $x(t)$ encodes the neuron's activity level. The delay τ represents the transmission time between output $x(t)$ and input.

Typically, many neurons are connected into a neural network so one has

$$y_i'(t) = -A_i y_i(t) + \sum_{j=1}^{n} W_{ij} \tanh(y_j(t - \tau_{ij})) + I_i(t) \qquad (1.13)$$

where y_i is the activity of the ith neuron, τ_{ij} is the delay in signal transmission between the jth neuron and ith neuron, W_{ij} is the weighting of the connection between the jth neuron and the ith neuron, I_i represents other inputs to the ith neuron. See van den Driessche, Wu, and Zou [76]. Wei and Ruan [79] show that oscillations may occur for the two-neuron system:

$$u_1'(t) = -u_1 + 2\tanh(u_2(t - \tau_{12})) \qquad (1.14)$$

$$u_2'(t) = -u_2 - 1.5\tanh(u_1(t - \tau_{21}))$$

Control of gene expression in cells is often modeled with time delays in equations of the form

$$x_1'(t) = g(x_n(t - r_n)) - \alpha_1 x_1(t) \qquad (1.15)$$

$$x_j'(t) = x_{j-1}(t - r_{j-1}) - \alpha_j x_j(t)$$

The gene is transcribed producing mRNA(x_1) which is translated into enzyme x_2 and it in turn produces another enzyme x_3 and so on. The end product x_n acts to repress the transcription of the gene $g' < 0$. Time delays are introduced to account for time involved in transcription, translation, and transport. The $\alpha_j > 0$ represent decay rates of the species. See [3, 32, 68] for more references on these equations. We consider (1.15) in Chapter 6.

Many delay differential equations arise in mathematical modeling of physiological processes. Volume 2 of Keener and Sneyd's book on the subject [47] is a good source for such examples.

Similar delayed feedback systems have been introduced to model the control of testosterone levels in the blood stream. Murray [61][chapter 6, section 6] introduces the model

$$R'(t) = f(T(t)) - b_1 R(t)$$
$$L'(t) = g_1 R(t) - b_2 L(t) \tag{1.16}$$
$$T'(t) = g_2 L(t - \tau) - b_3 T(t)$$

The hypothalmus secretes LHRH, R, which controls the release of LH, L, by the pituitary which controls the production of testosterone, T, in the gonads. The delay τ accounts for the blood circulation time in the body. $f(T)$ models the feedback on the production of R by testosterone T; it satisfies $f(0) > 0$ and $f' < 0$. Enciso and Sontag [30] correct an error in the analysis of this system in [61].

Ellermeyer [28] and Ellermeyer, Hendrix, and Ghoochan [29] introduce a delay in the standard bacterial growth model in a chemostat to obtain the model system

$$S'(t) = D(S_0 - S(t)) - \frac{v_m S(t)}{C_h + S(t)} x(t) \tag{1.17}$$
$$x'(t) = \exp(-D\tau) \frac{Y v_m S(t - \tau)}{C_h + S(t - \tau)} x(t - \tau) - Dx(t)$$

Here $S(t)$ denotes the substrate (food for bacteria) concentration, $x(t)$ is the biomass concentration of bacteria, and Y is a yield factor converting substrate to bacterial biomass (a unit of substrate produces Y units of bacteria). A chemostat can be viewed as a well-stirred vessel with fresh substrate at concentration S_0 poured in at rate D and unused nutrient and bacteria being drained out at the same rate D to keep the volume fixed. See [69] for more on chemostats. The delay τ reflects the assumption that whereas cellular absorption of substrate is assumed to be an instantaneous process, a resulting increase in microbial biomass reflecting assimilation is assumed to lag by a fixed amount of time τ. Experimental work in [29] found a delay of 20 minutes for a strain of E. coli. The argument for the bacterial growth term in (1.17) is as follows. During the brief time interval $[t - \tau, t - \tau + dt]$ an amount of substrate $Q_0 = f(S(t - \tau))x(t - \tau)dt$ is taken up by the cells in the chemostat, where $f(S) = v_m S / (C_h + S)$ is used for brevity. At time t this substrate will have been converted to bacterial biomass inside the cells, some of which will have drained out of the chemostat and therefore must not be counted. To see how much of Q_0 remains in the chemostat at time t, let $Q(s)$ denote the amount remaining in the chemostat at time $t - \tau + s$. As a constituent of the chemostat just like bacteria and substrate, it is subject to draining out according to $dQ/ds = -DQ$ with initial condition Q_0 at $s = 0$. Therefore at $s = \tau$, which corresponds to time t, $Q = Q_0 e^{-D\tau}$. Multiplying by Y gives that portion of the biomass created that remains in the chemostat at time t. This accounts for the factor $e^{-D\tau}$ in (1.17). This system is treated in detail in Chapter 5.

Culshaw and Ruan [22] modify a standard within-host HIV model to include a time delay between virus-cell contact and subsequent infection of the CD4+ T-cell. The model is given by

$$T'(t) = s - \mu_T T(t) + rT(t)(1 - \frac{T(t) + I(t)}{T_{max}})$$
$$-k_1 T(t)V(t)$$
$$I'(t) = k_1' T(t - \tau)V(t - \tau) - \mu_1 I(t) \tag{1.18}$$
$$V'(t) = N\mu_b I(t) - k_1 T(t)V(t) - \mu_V V(t)$$

where T denotes healthy (uninfected) T cells in the blood, I are the T cells infected by the HIV virus, and V is the virus level in the blood. In more recent work, Culshaw, Ruan, and Webb [23] cite evidence that in lymphatic tissue direct cell-to-cell transmission of HIV is the dominant mode of infection. If C denotes concentration of healthy cells and I is concentration of infected cells, they derive the model system:

$$C'(t) = rC(t)\left(1 - \frac{C(t) + I(t)}{C_M}\right) - k_I I(t)C(t)$$
$$I'(t) = k_I' \int_{-\infty}^{t} I(u)C(u)F(t - u)du - \mu_I I(t) \tag{1.19}$$

k_I'/k_I is the fraction of cells surviving the incubation period. The integral term is explained as follows. A cell that becomes productively infectious at time t was infected at a time u in the past with probability $F(u)$. This is best appreciated by making a change of variables in the integral. If $F(u) = \delta(u)$, the unit impulse function at $u = 0$, the system becomes an ODE. If $F(u) = \delta(u - \tau)$, it becomes a delay equation with discrete delay. Finally, if

$$F(u) = \frac{a^p u^{p-1}}{(p-1)!}e^{-au}$$

the delay is effectively infinite but centered at p/a. Infinite delays of this type are considered in Chapter 7.

Lenski and Levin [51] model phage (virus that attacks bacteria) growth on a bacterial host that consumes a limiting nutrient in a chemostat by the system

$$R'(t) = D(R_0 - R(t)) - f(R(t))S(t)$$
$$S'(t) = (f(R(t)) - D)S(t) - kS(t)P(t) \tag{1.20}$$
$$I'(t) = kS(t)P(t) - DI(t) - e^{-D\tau}kS(t - \tau)P(t - \tau)$$
$$P'(t) = -DP(t) - kS(t)P(t) + be^{-D\tau}kS(t - \tau)P(t - \tau)$$

R is the resource supporting bacterial growth, S is uninfected bacteria, I is phage-infected bacteria, and P is phage. Parameter b denotes the burst size, the number of phage progeny released following lysis of the host cell; k denotes the phage adsorption rate, R_0 is input nutrient supplied to bacteria, D is the dilution rate of the chemostat, and $f(R)$ is the specific growth rate of bacteria at resource level R. The delay τ accounts for the phage latent period, the time from binding to host cell, and subsequent host cell lysis. The factor $e^{-D\tau}$ appears for reasons analogous to those described above for (1.17). A recent mathematical analysis of the model was carried out by Beretta, Solimano, and Tang [7]. We devote Chapter 8 to a study of its dynamics.

A recent survey of predator-prey models with time delays by Ruan [64] is highly recommended.

H.-O. Walther [77] considers the problem of making a soft landing from inferred position and velocity data. Suppose you travel on a straight line, the u-axis, and desire to make a soft landing at the origin. For definiteness, suppose that you are on the positive u-axis. Then you want:

$$u(t) > 0,\ t < t_0 \le \infty,\ \lim_{t \uparrow t_0}(u(t), u'(t)) = (0,0) \qquad (1.21)$$

where t_0 is the time of landing. You want to devise a control mechanism that lets you take your current position and velocity and use it to compute the desired acceleration in order to achieve a soft landing. But how will you know your current position and velocity? You might send a signal that travels at speed c, reflects off the landing site $u = 0$, and returns. If it is sent at time $t - s$ and returns at time t, then

$$cs = u(t-s) + u(t) \qquad (1.22)$$

Of course, you don't know either $u(t)$ or $u(t - s)$ but you can measure s. If c is large then $u(t) \approx u(t - s)$ so a good estimate of your position at time t is

$$p(t) = \frac{cs}{2}$$

and your velocity can be estimated as $p'(t)$. Your problem then is to define the control function a such that (1.21) is achieved for a large set of initial data for the system

$$u' = v \qquad (1.23)$$
$$v' = a(p, p')$$

Clearly, you want $a(0,0) = 0$ but what else? Walther first considers the case where position and velocity are precisely known ($p = u, p' = v$) in the equations above and restricts attention to linear $a = \alpha u + \beta v$. Consideration of the resulting linear constant coefficient system and its eigenvalues motivates the choice $\beta < 0$ and $-\beta^2/4 < \alpha < 0$. See the article [77] for more on this problem. We include it here to point out the possibility that a delay may depend on the state of the system instead of being merely a constant. Such delays are called state-dependent delay and are a current focus of research [42].

Consider a pendulum that can move through an entire circle. Is it possible to stabilize the "straight-up" steady state? Consider the damped pendulum equation where we have added a delayed negative feedback restoring force to try to stabilize the straight-up equilibrium $\theta = \pi$:

$$ml\theta''(t) + k\theta'(t) + mg\sin(\theta(t)) = R(\pi - \theta(t-r)) \qquad (1.24)$$

where θ represents the (counterclockwise) angle from the straight-down position. Parameter m denotes the mass on the end of the pendulum of length l; g is the gravitational constant and k is a measure of the damping. Recall that when $R = 0$, the "down" steady state $\theta = 0$ is asymptotically stable if $k > 0$ and the "up" steady-state $\theta = \pi$ is an unstable saddle point. Can we choose the "gain" $R > 0$ so as to stabilize the up steady-state? The delay r reflects the time needed to measure θ and implement the stabilizing feedback. For more recent work on stabilizing the vertical position of an inverted pendulum see Landry et al. [49]. Equation (1.24) is studied in Chapter 6.

1.2 Some Terminology

We introduce here some commonly used terminology. If $x(t)$ is a function of time, then the delay r in the argument of $x(t-r)$ is called a *discrete delay* and delay differential equations such as (1.1) involving only discrete delays are said to be of discrete delay type or, more simply, to have discrete delays. A term like

$$\int_{t-r}^{t} k(t-s)x(s)ds = \int_{0}^{r} k(z)x(t-z)dz$$

where $0 \leq r \leq \infty$ is referred to as a *distributed delay* because it reflects a weighted average of delays $x(t-z)$. Distributed delays are thought to be more realistic but they are more difficult to work with and the kernel k may be difficult to estimate from data. Delay differential equations such as (1.4) containing these kinds of delays are said to have distributed delays.

If the delay kernel $k(u)$ in (1.4) (or in (1.2)) is identically zero for all $u > u_{max}$, then the delay represented by the integral term

$$\int_{-\infty}^{t} N_1(s)k(t-s)ds = \int_{0}^{\infty} N_1(t-u)k(u)du$$

is said to be a *bounded delay* because the integral only samples values of N_1 for a bounded set of past times $[t - u_{max}, t)$. Otherwise, we call the delay an *unbounded delay* or an *infinite delay*. Discrete delays are obviously bounded delays. Equation (1.19) has an unbounded distributed delay when

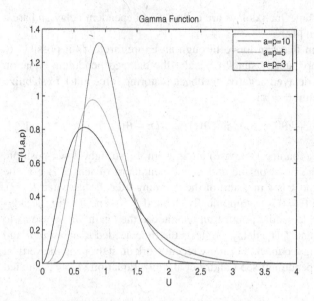

Fig. 1.1 $F(u,a,p)$ for $a = p = 3, 5, 10$. The mean delay is one.

$$F(u) = F(u,a,p) = \frac{a^p u^{p-1}}{(p-1)!} e^{-au}.$$

See Figure 1.1. Unbounded delays are more challenging to deal with because there are many possible classes of initial data one may use. These notes primarily treat equations with bounded delays. However, Chapter 7 is devoted to infinite delays.

There are many variations that may occur. For example, a discrete (or distributed) delay may be time-dependent such as $x(t - r(t))$ where $r(t) \geq 0$ is some prescribed function. It may also be a *state-dependent delay* such as $x(t - r(x(t)))$. This is the form of the delay defined implicitly by (1.22). These kinds of delays are especially challenging from the mathematical viewpoint.

What do we mean by a solution of an equation like (1.1)? If $N(t)$ is differentiable on $a \leq t < b$ and satisfies the equation on that interval then $N(t - r)$ must be defined for all such t. This means that N must be defined at least on $a - r \leq t \leq b$ although $N(t)$ need not be differentiable on $a - r \leq t \leq a$. This motivates the following definition. $N : [a - r, b) \to \mathbb{R}$ is a solution of (1.1) if it is continuous on its domain, differentiable on $[a, b)$ (right-differentiable at a), and satisfies the equation for $a \leq t < b$.

Most of the equations described above are of *autonomous* type inasmuch as the time t does not appear explicitly in the equation, (1.10) being the only exception. It is called a *nonautonomous* equation. Just as for ODEs, equations of autonomous type have the property that any time translate of a solution is again a solution. For example, the reader may easily verify that if $N(t)$ is a solution of (1.1) defined on

$a - r \leq t \leq b$, where $a < b$, and c is any real number then $n(t) = N(t - c)$ is a solution defined on $a + c - r \leq t \leq b + c$.

1.3 Solving Delay Equations Using a Computer

Anything written here is sure to be out of date soon. At the time of writing, the following were found to be useful.

A tutorial on solving delay equations using MATLAB by Shampine and Thompson can be found at the web site http://www.runet.edu/ thompson/webddes/

A list of software available for delay differential equations is maintained at the site http://twr.cs.kuleuven.be/research/software/delay/software.shtml

Among these is DDE–BIFTOOL v. 2.03, a MATLAB package for bifurcation analysis of delay differential equations. See [31].

Chapter 2
Delayed Negative Feedback: A Warm-Up

Abstract A great deal about delay differential equations can be learned by a study of its simplest representative, the linear delayed negative feedback equation. We use it to illustrate features common to delay differential equations, such as the tendency of delays to give rise to oscillations that can become undamped if delays are large. The obstructions to solving delay differential equations backwards in time are readily appreciated for this simple equation. It is an unpleasant fact that even for this simple linear equation, the stability of the trivial equilibrium requires an analysis of the roots of a transcendental equation. We show how the leading root of this transcendental equation signals the oscillation in solutions of the delay differential equation.

2.1 Preliminaries

The simplest delay differential equation is given by

$$u'(t) = -u(t - \tau) \tag{2.1}$$

where $\tau > 0$ is called the delay and the negative sign on the right indicates negative feedback. When $\tau = 0$, we recover the simple ODE

$$u'(t) = -u(t) \tag{2.2}$$

whose general solution, $u(t) = u(0)e^{-t}$, decays to zero.

If we prescribe $u(t)$ for $-\tau \leq t \leq 0$, then Equation (2.1) should have a unique solution for $t > 0$. Suppose we set

$$u(t) = 1, -\tau \leq t \leq 0 \tag{2.3}$$

as "initial data" for (2.1). Then, on the interval $0 \leq t \leq \tau$ the argument of u on the right side satisfies $t - \tau \leq 0$ so

H. Smith, *An Introduction to Delay Differential Equations with Applications to the Life Sciences*, 13
Texts in Applied Mathematics 57, DOI 10.1007/978-1-4419-7646-8_2,
© Springer Science+Business Media, LLC 2011

$$u'(t) = -u(t - \tau) = -1$$

and therefore

$$u(t) = u(0) + \int_0^t (-1)ds = 1 - t, 0 \le t \le \tau. \tag{2.4}$$

On $\tau \le t \le 2\tau$, we have $0 \le t - \tau \le \tau$ so by (2.4) we have

$$u'(t) = -u(t - \tau) = -[1 - (t - \tau)]$$

and thus

$$u(t) = u(\tau) + \int_\tau^t -[1 - (s - \tau)])ds$$

$$= 1 - \tau + [-s + \frac{1}{2}(s - \tau)^2|_{s=\tau}^{s=t}$$

$$= 1 - t + (t - \tau)^2/2, \tau \le t \le 2\tau. \tag{2.5}$$

In exercise 2.1, we ask the reader to verify that

$$u(t) = 1 + \sum_{k=1}^{n} (-1)^k \frac{[t - (k-1)\tau]^k}{k!}, (n-1)\tau \le t < n\tau, n \ge 1. \tag{2.6}$$

Thus, $u(t)$ is a polynomial of degree n on each subinterval of the form $[(n-1)\tau, n\tau)$. It follows that $u(t)$ is a smooth function, except at each $n\tau$, $n \ge 0$. The formulas (2.4), (2.5), and (2.6) imply that

(a) $u'(0-) = 0$ and $u'(0+) = -1$ so u' has a jump discontinuity at $t = 0$.
(b) $u''(\tau-) = 0$ and $u''(\tau+) = 1$ so u'' has a jump discontinuity at $t = \tau$.
(c) $u^{(n)}((n-1)\tau-) = 0$ and $u^{(n)}((n-1)\tau+) = (-1)^n$.
(d) u is n-times continuously differentiable on $((n-1)\tau, \infty)$ for $n \ge 0$.

Here, $u^{(j)}(s+)$ denotes the limit of the j-th derivative of u as $t \to s$, $t > s$ and $u^{(j)}(s-)$ denotes the limit as $t \to s$, $t < s$. A key point is that u gets smoother as t increases.

The procedure used to solve the initial-value problem (2.1) and (2.3) above is called the method of steps for obvious reasons.

Let's begin by exploring this solution numerically by using the MATLAB DDE23 package. We want to investigate the behavior of the solution on the interval $t > 0$ for different values of the delay τ. Figure 2.1 shows the results.

Notice that the case $\tau = 0.25$ looks very much like the solution of the ODE (2.2) with $u(0) = 1$, namely it decays to zero without "overshooting" zero, i.e., it does not oscillate. When $\tau = 0.6$ the solution oscillates. In fact, despite appearances, it repeatedly changes sign. One can prove that all solutions oscillate whenever $\tau > e^{-1}$. Why does $\tau > e^{-1} \approx 0.36$ result in oscillations? We answer this later in the chapter. As τ increases the oscillations appear to be more pronounced but still they are damped. That is, it appears that the amplitude is decreasing, at least until $\tau = 2$ where now the amplitude grows.

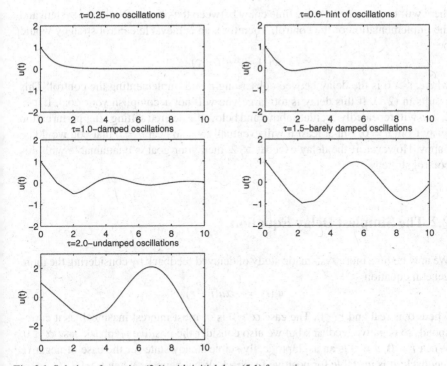

Fig. 2.1 Solution of equation (2.1) with initial data (5.1) for various τ

The reader will notice that $u \equiv 0$ is a solution of the delay differential equation (2.1); we call it a steady-state solution. We prove that it is stable when $\tau < \pi/2$ and unstable when $\tau > \pi/2 \approx 1.58$.

Equation (2.1) provides a simple illustration of what can go wrong in a negative feedback system with delays. Say you want to maintain a certain quantity u at the value $u = 0$. Imagine u satisfies a simple equation such as

$$u'(t) = c(t)$$

where you can prescribe the control $c(t)$ in order to accomplish your objective of maintaining u near zero. The system, of course, is subject to unexpected perturbations so you better be ready to handle these. If you observe that $u(t) > 0$ ($u(t) < 0$), you will want to choose $c(t) < 0$ ($c(t) > 0$). You might want to choose

$$c(t) = -\alpha u(t), \alpha > 0$$

because then no matter what value u is at time t_0, it will return to $u = 0$. For simplicity here, we take the "gain" parameter $\alpha = 1$. But consider how you would implement this feedback law. You must observe the system at time t to find $u(t)$ and then immediately respond with the control $c(t) = -u(t)$. This is clearly impossible;

there will necessarily be some time delay between the observation of the system and
the implementation of the control. Therefore, an achievable control strategy would
be

$$c(t) = -u(t - \tau)$$

where $\tau > 0$ is the delay between observing u and implementing the control. This
results in (2.1). If this delay is too large you will not accomplish your goal. If $\tau >
e^{-1}$, u will repeatedly oscillate above and below the desired setting after perturbation
away from $u = 0$. But at least it will eventually get so near $u = 0$ that you would be
happy. However, if the delay exceeds $\pi/2$, then your goal is unattainable with this
control strategy.

2.2 The Simplest Delay Equation

We now begin a more systematic study of delayed feedback by considering the more
general equation

$$u'(t) = -\alpha u(t - r) \tag{2.7}$$

where α is real and $r \geq 0$. The case $\alpha > 0$ is of most interest inasmuch as it corresponds to negative feedback but we also consider the positive feedback case $\alpha < 0$.
When $r = 0$, $u = 0$ is an asymptotically stable steady state for the case of negative
feedback; it is unstable for positive feedback. What happens when $r > 0$?

Scaling can reduce the number of parameters and produce simpler equations.
Here, there are two natural choices. By a scaling of time: $\tau = \eta t$, $\eta > 0$, the equation
for $U(\tau) = u(t)$ becomes:

$$\frac{dU}{d\tau} = \eta^{-1}\frac{du}{dt} = -\alpha\eta^{-1}u(t-r) = -\alpha\eta^{-1}U(\tau - r\eta)$$

If we take $\eta = 1/r$ and $\beta = \alpha r$ we get

$$\frac{dU}{d\tau} = -\beta U(\tau - 1). \tag{2.8}$$

We could let $\eta = |\alpha|$ and $s = r|\alpha|$, resulting in

$$\frac{dU}{d\tau} = \pm U(\tau - s)$$

where the sign is determined by that of α. Both forms are attractive but we choose
to work with (2.8).

In order to determine stability of the trivial solution, we proceed exactly as for
ODEs. That is, we seek (complex) values of λ such that $U(\tau) = \exp(\lambda\tau)$ is a solution of (2.8).

It is convenient to introduce the linear operator, defined on the differentiable
functions, by

$$L(U) = \frac{dU}{d\tau} + \beta U(\tau - 1)$$

Then

$$L(e^{\lambda\tau}) = \lambda e^{\lambda\tau} + \beta e^{\lambda(\tau-1)} = e^{\lambda\tau}[\lambda + \beta e^{-\lambda}] \qquad (2.9)$$

Clearly, we have $L(e^{\lambda\tau}) \equiv 0$ (the zero function!) if and only if λ is a root of the *characteristic equation*

$$h(\lambda) \equiv \lambda + \beta e^{-\lambda} = 0. \qquad (2.10)$$

We call $\lambda \in \mathbb{C}$ a root of (2.10) of order l, where $l \geq 1$, if

$$h(\lambda) = h'(\lambda) = h''(\lambda) = \cdots = h^{(l-1)}(\lambda) = 0, \; h^{(l)}(\lambda) \neq 0.$$

Lemma 2.1. $\tau^j e^{\lambda\tau}$, $j = 0, 1, \ldots, k$ *are solutions of* (2.8) *if and only if* λ *is a root of order at least* $k + 1$ *of h.*

Proof. Differentiating (2.9) k times with respect to λ and using that this kth-derivative commutes with L we find, by Leibniz' rule for the derivative of a product, that

$$L(\tau^k e^{\lambda\tau}) = (\frac{\partial}{\partial\lambda})^k[e^{\lambda\tau}h(\lambda)] = e^{\lambda\tau}[\sum_{j=0}^{k} C_j^k h^{(j)}(\lambda)\tau^{k-j}]$$

where $C_j^k = k!/j!(k-j)!$ are the binomial coefficients. The result follows immediately from this observation. $\square$

Alternatively, (2.10) can be obtained by using the Laplace transform.

Based on our experience with ODEs, we have a right to expect that the trivial solution is asymptotically stable if $\Re(\lambda) < 0$ for all roots λ of the characteristic equation and that it is unstable if there is a root with positive real part. We assume this now; it is proved later.

As h is an analytic function of the complex variable λ it has the following elementary properties. See Appendix A.

(A) The set of roots can have no accumulation point in $\mathbb{C}$; therefore, for each $R > 0$, the set of roots satisfying $|\lambda| \leq R$ is finite. It follows that the set of roots is a countable (possibly finite) set.
(B) If the set of roots is infinite, denoted by $\{\lambda_n\}_{n=1}^{\infty}$, then $|\lambda_n| \to \infty$. Because $|\beta|e^{-\Re(\lambda_n)} = |\lambda_n|$, it follows that $\Re(\lambda_n) \to -\infty$. Consequently, for each $a \in \mathbb{R}$, $\Re(\lambda) \geq a$ for at most finitely many roots.
(C) If λ is a root, then it is a root of finite order.
(D) If λ is a root, so is its conjugate $\overline{\lambda}$.

Now let's focus attention on the characteristic equation (2.10). Letting $\lambda = x + iy$ and considering real and imaginary parts, we get the system

$$x = -\beta e^{-x} \cos(y) \qquad (2.11)$$
$$y = \beta e^{-x} \sin(y)$$

We begin by considering real roots of (2.10).

Lemma 2.2. *The following hold.*

1. *If $\beta < 0$, then there is exactly one real root and it is positive.*
2. *If $0 < \beta < e^{-1}$, then there are exactly two real roots $x_1 < x_2$, both negative. $x_1 \to -\infty$ and $x_2 \to 0$ as $\beta \to 0$.*
3. *If $\beta = e^{-1}$, then there is a single real root of order two, namely $\lambda = -1$.*
4. *If $\beta > e^{-1}$, then there are no real roots.*

The proof is left to the exercises.

The next result summarizes important information concerning the roots in the case $\beta > 0$.

Proposition 2.1 *The following hold for (2.10).*

1. *If $0 < \beta < \pi/2$, then there exists $\delta > 0$ such that $\Re(\lambda) \leq -\delta$ for all roots.*
2. *If $\beta = \pi/2$, then $\lambda = \pm i\pi/2$ are roots of order one.*
3. *If $\beta > \pi/2$, there are roots $\lambda = x \pm iy$ with $x > 0, y \in (\pi/2, \pi)$.*

The following immediate corollary of Proposition 2.1 follows once we establish the expected result concerning the relation of the roots of (2.10) and the stability of the zero solution of (2.7).

Corollary 2.2 *The following hold for (2.7).*

1. *If $\alpha < 0$ then $u = 0$ is unstable.*
2. *If $0 < r\alpha < \pi/2$, $u = 0$ is asymptotically stable.*
3. *If $r\alpha = \pi/2$, $u = \sin(\pi\tau/2), \cos(\pi\tau/2)$ are solutions.*
4. *If $r\alpha > \pi/2$, $u = 0$ is unstable.*

Figure 2.2 depicts the stability region in the (r, α)-plane and Figure 2.3 shows two simulations, both with $\alpha = 1$. The delay r is just below $\pi/2$ for the right graph and just above it for the left graph.

Proof. **Proof of Proposition 2.1:** Because $\beta > 0$, if there is a root $x + iy$ with $x \geq 0$ and $y > 0$ of (2.11) then $\cos(y) \leq 0 < \sin(y)$ so $y \in S \equiv \cup_{n=0}^{\infty}\{[\pi/2, \pi) + 2n\pi\}$. Furthermore,

$$\frac{\sin(y)}{y} = \frac{e^x}{\beta}$$

must hold. As

$$\frac{d}{dy}\frac{\sin(y)}{y} = \frac{y\cos(y) - \sin(y)}{y^2} < 0, y \in S$$

and $\sin(y)/y = 2/\pi$ when $y = \pi/2$, we conclude that $\sin(y)/y \leq 2/\pi$ for $y \in S$ even though S is disconnected (check this). Therefore,

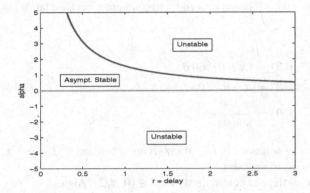

Fig. 2.2 The stability region in the (r, α)-plane for (2.7).

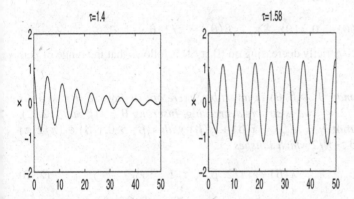

Fig. 2.3 Simulation of (2.7) for $\alpha = 1$, $r = \tau$ initial data equal one.

$$\frac{1}{\beta} \leq \frac{e^x}{\beta} = \frac{\sin(y)}{y} \leq \frac{2}{\pi},$$

so it follows that $\beta \geq \pi/2$. Thus, if $\beta < \pi/2$ then $\Re(\lambda) < 0$ for every root λ. This and the last assertion in (B) above proves (1).

Let's now turn to the final assertion (3). If we write our root as $\lambda = re^{i\theta}$, then (2.10) becomes

$$r[\cos(\theta - \pi) + i\sin(\theta - \pi)] = \beta e^{-x}[\cos(-y) + i\sin(-y)]$$

Equivalently,

$$r = \beta e^{-x} \text{ and } \theta - \pi = -y + 2k\pi$$

for some integer k. Let's search for a root in the first quadrant on the ray through the origin making angle $\theta \in (0, \pi/2)$ with the positive x-axis. Then, taking $k = 0$,

$y(\theta) = \pi - \theta > 0$ and so $x(\theta) > 0$ is determined by trigonometry because $\tan(\theta) = y/x$ (draw a picture).

We claim that:

$$x(\theta) = (\pi - \theta)\cot(\theta)$$
$$y(\theta) = \pi - \theta, \quad 0 < \theta < \pi/2 \tag{2.12}$$
$$\beta(\theta) = \frac{\pi - \theta}{\sin(\theta)}e^{x(\theta)}$$

is a one-parameter family of solutions of (2.10) satisfying $x > 0$ and $\pi/2 < y < \pi$. The proof of the claim is left to the exercises.

Clearly, $x(\theta), y(\theta), \beta(\theta)$ depend continuously on $\theta \in (0, \pi/2)$. Also,

$$x(\theta) \to +\infty, \ y(\theta) \to \pi, \ \beta(\theta) \to +\infty, \ \theta \to 0$$

and

$$x(\theta) \to 0, \ y(\theta) \to \pi/2, \ \beta(\theta) \to \pi/2, \ \theta \to \pi/2.$$

Inasmuch as $\beta(\theta)$ is strictly decreasing on $(0, \pi/2)$ it follows that the range of β is $(\pi/2, \infty)$. $\square$

Remark 2.3 *The map* $\theta \to \beta(\theta)$ *is invertible, therefore its inverse* $\theta = k(\beta)$ *is defined for* $\beta \in (\pi/2, \infty)$ *and is strictly decreasing. Inserting* $\theta = k(\beta)$ *into (2.12), we have found a root* $\lambda = \lambda(\beta) = x(\beta) + iy(\beta)$ *with* $x(\beta) > 0, y(\beta) \in (\pi/2, \pi)$ *corresponding to* $\beta > \pi/2$ *which satisfies*

$$\lambda(\beta) \to i\pi/2, \ \beta \searrow \pi/2 \tag{2.13}$$

Remark 2.4 *It's easy to see from (2.11) that for* $\beta = \pi/2 + 2n\pi, \ n = 0, 1, 2, \cdots,$ *then* $i[\pi/2 + 2n\pi]$ *is a root.*

2.3 Oscillation of Solutions

We can note that unlike the undelayed negative feedback case delayed negative feedback can result in oscillatory solutions. Let's be precise about what we mean. If $x(t)$ is a solution of (2.7) defined for $t \geq s$ for some real s, we say it is oscillatory if it has arbitrarily large zeros: for every $t_0 > s$ there exists $t_1 > t_0$ such that $x(t_1) = 0$; otherwise we say that the solution is nonoscillatory. The following is Theorem 2.1.3 in [39]; See also Theorem 1.5.1 in [37].

Theorem 2.5 *For every real* α *and* $r > 0$ *the following are equivalent.*

(a) Every solution of (2.7) is oscillatory.
(b) $r\alpha > 1/e$.

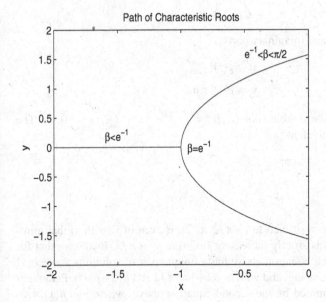

Fig. 2.4 Path of key characteristic root of (2.10) as a function of β. As β increases the root(s) move right.

Recall that by Lemma 2.2, Theorem 2.5 (b) is equivalent to there being no real roots of the characteristic equation. Therefore implication (a) implies (b) is obvious; the converse is not hard to prove. See [39], an excellent reference for the oscillatory behavior of delay differential equations. What happens at $\beta = r\alpha = e^{-1}$ that suggests solutions oscillate for larger values of β? This value of β corresponds to the double root $\lambda = -1$. See Figure 2.4 where we trace the path of the largest real root as β increases from $\beta \ll e^{-1}$ to $\beta = \pi/2$. A complex conjugate pair of roots bifurcates from the double root. This leads to oscillations.

A closer look at case (3) of Lemma 2.2 shows the following.

Lemma 2.3. *For each $\beta \in (e^{-1}, \pi/2)$ there is a complex conjugate pair of roots $\lambda = x \pm iy$ of (2.10) satisfying*

$$-1 < x < 0, 0 < y < \pi/2$$

This family of roots (λ, β) can be parameterized by x:

$$\beta = \beta(x), y = h(x)$$

where $\beta(x), h(x)$ are smooth, positive, monotonic increasing functions such that as $x \searrow -1$, $\lambda \to -1$, and $\beta \to e^{-1}$, and as $x \nearrow 0$, $\lambda \to (\pi/2)i$, and $\beta \to \pi/2$.

Proof. We start by changing variables $\lambda = -1 - z$, $\beta = e^{-1-\mu}$ in (2.10) so it becomes

$$z + 1 = e^{z-\mu}$$

or, if $z = x + iy$, in real and imaginary parts:

$$x + 1 = e^{x-\mu} \cos y$$
$$y = e^{x-\mu} \sin y$$

We have transformed the double root $(\lambda, \beta) = (-1, e^{-1})$ to $(z, \mu) = (0,0)$. The equations above are equivalent to

$$\frac{\tan y}{y} = \frac{1}{x+1}$$
$$e^{x-\mu} = ((x+1)^2 + y^2)^{1/2}$$

Function $g(y) = \tan y / y$, restricted to $(-\pi/2, \pi/2)$, is even in y, with global minimum of 1 at $y = 0$, and is strictly increasing (to ∞) as $y \to \pi/2$. It follows that for each $x \in (-1, 0)$, the first equation above has a unique pair of solutions $y = \pm h(x)$ where h is positive, decreasing, and with $h(-1+) = \pi/2$ and $h(0-) = 0$. Parameter $\mu = \mu(x)$ is now determined by the second equation above, where $y = h(x)$. Observe that $\mu(-1+) = -1 - \ln(\pi/2)$ and $\mu(0-) = 0$. It remains to show that $\mu(x)$ is strictly increasing. An alternative relation satisfied by $\mu(x)$ is

$$e^{\mu(x)} = e^x \frac{\sin(h(x))}{h(x)}$$

Because $\sin(y)/y$ is strictly decreasing on $(0, \pi/2)$ and $h(x)$ is strictly decreasing, it follows that their composition is strictly increasing and therefore so is $e^{\mu(x)}$. The proof is completed by putting the results above together with the change of variables. □

Lemma 2.3 and Proposition 2.1 imply that when $\beta = r\alpha > e^{-1}$, there is an oscillatory solution

$$u(t) = e^{(x+iy)t/r}$$

of (2.7). It decays to zero if $\beta \in (e^{-1}, \pi/2)$, and becomes unbounded when $\beta > \pi/2$. This explains the magic number e^{-1} appearing in Theorem 2.5. Out of the double root $\lambda = -1$ at $\beta = e^{-1}$ is born a complex conjugate pair of roots that lead first to damped oscillation and then to undamped oscillation.

2.4 Solutions Backward in Time

ODEs can be solved backward in time as well as forward but delay equations are fundamentally different in this respect. As we show, for some initial data solutions in backward time exist but not for others. We have no intention here of treating the general case but rather just to see what the problems are.

Consider the initial-value problem treated in Section 1,

$$u'(t) = -u(t - \tau) \tag{2.14}$$
$$u(t) = 1, -\tau \le t \le 0$$

There, we found a solution $\hat{u}(t)$ defined for $t \ge -\tau$.

If we replace t by $t + \tau$ in the equation, we may write the equation as

$$u(t) = -u'(t + \tau) \tag{2.15}$$

Roughly, to solve for u in the past we must differentiate u in the future. Proceeding by steps, starting with $-2\tau \le t < -\tau$, and so forth leads to an extension of $\hat{u}$:

$$\hat{u}(t) = 0, t < -\tau$$

The discontinuity of $\hat{u}$ at $t = -\tau$ means that our function $\hat{u} : \mathbb{R} \to \mathbb{R}$, defined by the method of steps, does not satisfy the definition of a solution given in Chapter 1, which required continuity. But maybe that definition is too restrictive. A more serious problem is that $\hat{u}$ is not differentiable at $t = 0$ where its left-hand derivative 0 differs from its right-hand derivative -1. Still, maybe we should overlook this and not require differentiability at every point.

If instead we take initial data $u(t) = 0$, $-\tau \le t \le 0$, there is no problem: $u(t) = 0$, $t \in \mathbb{R}$ is our solution. Of course, $u = 0$ is an equilibrium (constant) solution, it is special. In fact, it is easy to see that the initial-value problem

$$u'(t) = -u(t - \tau), t > s$$
$$u(t) = 0, s - \tau \le t \le s$$

has the solution $u(t) = 0$, $t \in \mathbb{R}$. But consider the implication of this observation for our backward "solution" $\hat{u}$ of the initial-value problem (2.14). Because it satisfies $\hat{u}(t) = 0$, $t < -\tau$, obviously $\hat{u}(t) = 0$ on $-3\tau \le t \le -2\tau$, so we conclude that $u(t) = 0$, $t \in \mathbb{R}$ is another solution of the corresponding initial-value problem given by the initial data $u(t) = 0$ on $-3\tau \le t \le -2\tau$. Therefore, if we accept the backward "solution" $\hat{u}$ then we are forced to accept the nonuniqueness of solutions of initial-value problems. This is a strong argument against defining the concept of "solution" so as to allow $\hat{u}$ to be one.

It seems obvious that given any continuous initial data $\phi : [-\tau, 0] \to \mathbb{R}$, we can use the method of steps to solve the initial-value problem

$$u'(t) = -u(t - \tau) \tag{2.16}$$
$$u(t) = \phi(t), -\tau \le t \le 0$$

We prove this fact in the next chapter. There exist continuous functions ϕ that are not differentiable at any point of $-\tau \le t \le 0$. Clearly, we cannot use (2.15) to find a backward-in-time solution inasmuch as ϕ is not differentiable.

Exercises

Exercise 2.1. Verify (2.6).

Exercise 2.2. Show that if λ is a root of (2.10) then so is its conjugate $\bar{\lambda}$.

Exercise 2.3. Show that all roots of (2.10) have order one except when $\beta = e^{-1}$ when $\lambda = -1$ has order two.

Exercise 2.4. Prove Lemma 2.2.

Exercise 2.5. Prove the claim in the proof of Corollary 2.2 by inserting into (2.11). Show that $x(\theta)$ and $\beta(\theta)$ are strictly decreasing in $\theta \in (0, \pi/2)$.

Chapter 3
Existence of Solutions

Abstract Existence and uniqueness of solutions of discrete–delay differential equations is established by the method of steps, appealing to classical ODE results. More general delay equations require a more general framework for existence and uniqueness. This includes some peculiar notation endemic to the subject and the identification of the appropriate state space for delay equations. Solutions either extend to the entire half–line or blowup in finite time. Applications to biology require that solutions that start positive, stay positive in the future. Differential inequalities involving delays are important tools with which to bound solutions. All these topics are taken up here, including basic stability definitions

3.1 The Method of Steps for Discrete Delay Equations

Now let's consider the nonlinear delay differential equation

$$x'(t) = f(t, x(t), x(t - r)) \qquad (3.1)$$

with a single delay $r > 0$. Assume that $f(t, x, y)$ and $f_x(t, x, y)$ are continuous on $\mathbb{R}^3$. Let $s \in \mathbb{R}$ be given and let $\phi : [s - r, s] \to \mathbb{R}$ be continuous. We seek a solution $x(t)$ of (3.1) satisfying

$$x(t) = \phi(t), s - r \le t \le s \qquad (3.2)$$

and satisfying (3.1) on $s \le t < s + \sigma$ for some $\sigma > 0$. Note that we must interpret $x'(s)$ as the right-hand derivative at s.

Equation (3.1) can be solved by the *method of steps* as follows. For $s \le t \le s + r$, $x(t)$ must satisfy the initial-value problem for the ODE:

$$y'(t) = f(t, y(t), \phi(t - r)), y(s) = \phi(s), s \le t \le s + r$$

As $g(t, y) \equiv f(t, y, \phi(t - r))$ and $g_y(t, y)$ are continuous, a local solution of this ODE is guaranteed by standard results from ODE theory [10, 40]. If this local solution

H. Smith, *An Introduction to Delay Differential Equations with Applications to the Life Sciences*, Texts in Applied Mathematics 57, DOI 10.1007/978-1-4419-7646-8_3,
© Springer Science+Business Media, LLC 2011

$x(t)$ exists for the entire interval $s \leq t \leq s+r$, then our solution $x(t)$ is defined so far on $[s-r, s+r]$ and we may repeat the above argument to extend our solution still farther to the right. Indeed, for $s+r \leq t \leq s+2r$, a solution $x(t)$ of (3.1)-(3.2) must satisfy the initial value-problem for the ODE:

$$y'(t) = f(t, y(t), x(t-r)), y(s+r) = x(s+r), s+r \leq t \leq s+2r$$

Again, standard existence results for such problems guarantee the existence of a unique solution, which we call $x(t)$, defined on some subinterval $[s+r, \sigma) \subset [s+r, s+2r]$, possibly the entire interval. Of course, $x(t)$, now defined on $[s-r, \sigma)$ where $\sigma > s+r$, is a solution of (3.1)-(3.2). If the solution exists on the entire interval $[s+r, s+2r]$ then we may again repeat the process to extend the solution to $[s+2r, s+3r]$, or some subinterval of this interval.

Theorem 3.1 *Let $f(t,x,y)$ and $f_x(t,x,y)$ be continuous on $\mathbb{R}^3$, $s \in \mathbb{R}$, and let $\phi : [s-r,s] \to \mathbb{R}$ be continuous. Then there exists $\sigma > s$ and a unique solution of the initial-value problem (3.1)-(3.2) on $[s-r, \sigma]$.*

The proof is left to the exercises.

Theorem 3.1 provides only a local solution of (3.1)-(3.2). Just as for ODEs, we can often, but not always, extend this solution to be defined for all $t \geq s$. We use the notation $[s-r, \sigma)$ to denote either the open interval $[s-r, \sigma)$ or the closed interval $[s-r, \sigma]$. The uniqueness assertion of Theorem 3.1 implies that if $x : [s-r, \sigma) \to \mathbb{R}$ and $\hat{x} : [s-r, \rho) \to \mathbb{R}$ are two solutions of (3.1)-(3.2), then $x(t) = \hat{x}(t)$ for all t such that both are defined. Alternatively, this can be proved directly using Gronwall's inequality. If $[s-r, \sigma) \subset [s-r, \rho)$, we say that $\hat{x}$ is an *extension*, or *continuation*, of x and write $x \subset \hat{x}$. This defines a partial order relation on the (linearly ordered) set of all solutions of (3.1)-(3.2). Zorn's lemma [34] can then be used to establish the existence of a unique maximally defined solution (one for which there are no extensions) $x : [s-r, \sigma) \to \mathbb{R}$ just as in [10, 40]. We call such a solution a *noncontinuable solution* because it cannot be continued (extended) to a larger interval. This noncontinuable solution is necessarily defined on an interval that is open on the right because if $\sigma < \infty$ and if x were a solution on $[s-r, \sigma]$ then we could use Theorem 3.1 to obtain an extension of x to a larger interval, contradicting that x has no extension.

The next result shows that if a noncontinuable solution is not defined for all $t \geq s-r$, then it must "blow up" as $t \to \sigma$. This is the same conclusion as for the ODE theory [10, 40]; in fact that theory proves the result.

Theorem 3.2 *Let f satisfy the hypotheses of Theorem 3.1 and let $x : [s-r, \sigma) \to \mathbb{R}$ be the noncontinuable solution of the initial-value problem (3.1)-(3.2). If $\sigma < \infty$ then*

$$\lim_{t \to \sigma-} |x(t)| = \infty.$$

Proof. Let $x : [s-r, \sigma) \to \mathbb{R}$ be the noncontinuable solution and suppose that $\sigma < \infty$. Then $s + jr < \sigma \leq s + (j+1)r$ for some $j \in \{0, 1, 2, \ldots\}$ so the restriction of $x(t)$ to the interval $[s + jr, \sigma)$ is necessarily the noncontinuable solution of the initial-value problem for the ODE:

$$y' = f(t, y(t), x(t - r)), y(s + jr) = x(s + jr)$$

inasmuch as any extension of it would give an extension of $x(t)$ as a solution of (3.1)-(3.2). But then the result follows from the continuation theorem for ODEs [10, 40]. □

Remark 3.3 *Theorems 3.1 and 3.2 extend immediately to the case that $x \in \mathbb{R}^n$ and $f : \mathbb{R} \times \mathbb{R}^n \times \mathbb{R}^n \to \mathbb{R}^n$ with almost no change in proof; it also extends to multiple discrete delays $r_0 < r_1 < \cdots < r_m$ where $f = f(t, y(t), y(t - r_0), y(t - r_1), \ldots, y(t - r_m))$ with very little change. In Section 3.3, we treat such systems.*

3.2 Positivity of Solutions

Most delay differential equations that arise in population dynamics and epidemiology model intrinsically nonnegative quantities. Therefore it is important to establish that nonnegative initial data give rise to nonnegative solutions. Consider the simple example

$$x'(t) = y(t) - x(t - r)$$
$$y'(t) = y(t)$$

If $r = 0$ so there is no delay, then solutions corresponding to nonnegative initial data remain nonnegative in the future. This is easiest to see by applying the variation of constants formula to the first equation

$$x(t) = x(0)e^{-t} + e^{-t} \int_0^t e^s y(s) ds$$

Because $y(t) \geq 0$ and $x(0) \geq 0$, it follows that $x(t) \geq 0$, $t \geq 0$. However, if $r > 0$, then nonnegativity of solutions corresponding to nonnegative initial data fails, as easily seen by taking $y(0) = 0$ and $x(s) = -s/r$, $-r \leq s \leq 0$. Because $x(0) = 0$ and $x'(0) = -1$, the solution immediately becomes negative yet the initial data are nonnegative.

Given $x \in \mathbb{R}^n$, we write $x \geq 0$ when $x_i \geq 0$, $1 \leq i \leq n$; similar notation is used for $x \leq 0$. Let $\mathbb{R}^n_+$ denote the set of vectors $x \in \mathbb{R}^n$ such that $x \geq 0$.

Theorem 3.4 *Suppose that $f : \mathbb{R} \times \mathbb{R}^n_+ \times \mathbb{R}^n_+ \to \mathbb{R}^n$ satisfies the hypotheses of Theorem 3.1, Remark 3.3, and*

$$\forall i, t, \forall x, y \in \mathbb{R}^n_+ : x_i = 0 \Rightarrow f_i(t, x, y) \geq 0 \tag{3.3}$$

If the initial data ϕ in (3.2) satisfy $\phi \geq 0$, then the corresponding solution $x(t)$ of (3.1) satisfies $x(t) \geq 0$ for all $t \geq s$ where it is defined.

Proof. Recall that the analogous result for ODEs $x' = g(t,x)$ where $g : \mathbb{R} \times \mathbb{R}_+^n \to \mathbb{R}^n$ requires g to satisfy $g_i(t,x) \geq 0$ whenever $x \in \mathbb{R}_+^n$ satisfies $x_i = 0$. See Proposition B.7 in [69].

Let $x(t)$ denote the solution of (3.1) corresponding to the nonnegative initial data (3.2). On the interval $s \leq t \leq s+r$, $x(t)$ satisfies the ODE $x'(t) = g(t,x(t))$ where $g(t,x) = f(t,x,\phi(t-r))$. g is easily seen to satisfy the conditions in the previous paragraph and hence $x(t) \geq 0$ on the interval $s \leq t \leq s+r$. Now one just repeats the argument using the method of steps. $\square$

Remark 3.5 *If $f_i(s,x,y) < 0$ for some i,s and some $x,y \in \mathbb{R}_+^n$ with $x_i = 0$ then for initial data given by the nonnegative continuous function*

$$\phi(\eta) = [1 - \frac{(s-\eta)}{r}]x + \frac{(s-\eta)}{r}y, s-r \leq \eta \leq s$$

the corresponding solution of (3.1) satisfies $x_i(s) = 0$ and $x_i'(s) < 0$ so $x_i(t) < 0$ for all $t > s$ sufficiently near $t = s$. Thus (3.3) is a necessary condition for generic positivity preservation.

As an example, consider the discrete-delay predator-prey model

$$N_1'(t) = N_1(t)[b - N_2(t)] \tag{3.4}$$
$$N_2'(t) = N_2(t)[-c + N_1(t-r)]$$

where $b,c > 0$ and delay $r > 0$ reflects a delay in assimilation of consumed prey. We specify nonnegative initial conditions:

$$N_1(t) = \phi_1(t), -r \leq t \leq 0 \tag{3.5}$$
$$N_2(0) = N_2^0$$

If we identify $x = (x_1, x_2)$ where $x_i = N_i(t)$ and $y = (y_1, y_2)$ where $y_i = N_i(t-r)$, then our system has the form (3.1) where

$$f(t,x,y) = (x_1[b - x_2], x_2[-c + y_1])$$

Obviously, f, f_x exist and are continuous so Theorem 3.2 implies that (3.4)-(3.5) has a unique noncontinuable solution defined on some interval $[-r, \sigma)$ where $\sigma > 0$.

Let's first verify that this solution has nonnegative components. If $x, y \geq 0$ and $x_i = 0$ for some i, then $f_i(t,x,y) = 0$ so Theorem 3.4 implies that $N_i(t) \geq 0$, $0 \leq t < \sigma$, $i = 1, 2$ because we assume that $N_2^0 \geq 0$ and $\phi_1 \geq 0$.

Now we provide (crude) estimates that ensure that $\sigma = +\infty$. First, we note that as $N_2(t) \geq 0$ it follows that $N_1'(t) \leq bN_1(t)$, which implies that $N_1(t) \leq \phi_1(0)e^{bt}$, $0 \leq t < \sigma$. Now $N_1(t-r) \leq M_1$, $0 \leq t \leq r$, where $M_1 = \max \phi_1$, and $N_1(t-r) \leq \phi_1(0)e^{b(t-r)}$, $r \leq t$ so $N_1(t-r) \leq M_1 e^{bt}$, $t \geq 0$. Consequently, $N_2'(t) \leq M_1 N_2(t)e^{bt}$, $t \geq 0$. Integrating gives

$$N_2(t) \leq N_2^0 \exp\left(\int_0^t M_1 e^{bs} ds \right) = N_2^0 \exp\left(\frac{M_1}{b}(e^{bt} - 1) \right), 0 \leq t < \sigma$$

These crude estimates ensure that $\sigma = +\infty$ because if $\sigma < +\infty$, then $|(N_1(t), N_2(t))|$ becomes infinite as $t \to \sigma$ by Theorem 3.2, contradicting the estimates.

Comparison principles and differential inequalities are extremely useful. Below is an elementary but useful example.

Theorem 3.6 *Let* $f : \mathbb{R}^3 \to \mathbb{R}$ *satisfy the hypotheses of Theorem 3.1 and suppose that* $f(t,x,y)$ *is nondecreasing in* y. *Let* $x(t)$ *be the solution of* (3.1) *satisfying* (3.2) *on an interval* $I = [s, s+b]$ *for some* $b > 0$. *Let* $\bar{x}(t)$ *be continuous on* $[s-r,s] \cup I$, *differentiable on* I, *and satisfy*

$$\bar{x}'(t) \geq f(t, \bar{x}(t), \bar{x}(t-r)), t \in I \qquad (3.6)$$
$$\bar{x}(t) \geq \phi(t), s - r \leq t \leq s$$

Then $x(t) \leq \bar{x}(t)$, $t \in I$. *A parallel result holds with all inequalities reversed.*

Proof. The simplest proof uses the method of steps and the corresponding differential inequality results for ordinary differential equations. See, for example, Theorem 3.2 of Chapter 4 in [78]. Below, we give another proof.

First suppose that both inequalities in (3.6) are strict. We claim that $x(t) < \bar{x}(t)$, $t \in I$. If false there would exist $t_0 > s$ in I such that $x(t) < \bar{x}(t)$, $s \leq t < t_0$ and $x(t_0) = \bar{x}(t_0)$. It follows that $x'(t_0) \geq \bar{x}'(t_0)$. But

$$x'(t_0) = f(t_0, x(t_0), x(t_0-r)) \leq f(t_0, \bar{x}(t_0), \bar{x}(t_0-r)) < \bar{x}'(t_0)$$

because $x(t_0 - r) < \bar{x}(t_0 - r)$ and $x(t_0) = \bar{x}(t_0)$. This contradiction proves the result in this case.

For the general case, let $\varepsilon > 0$ and $x_\varepsilon(t)$ be the solution of $x'(t) = f_\varepsilon(t, x(t), x(t-r)) = f(t, x(t), x(t-r)) - \varepsilon$ corresponding to initial data $x_\varepsilon(t) = \phi(t) - \varepsilon$, $t \in [s-r, s]$. By the results of the previous paragraph, we may conclude that $x_\varepsilon(t) < \bar{x}(t)$ for all $t \in I$ for which $x_\varepsilon(t)$ is defined. It can be shown that for sufficiently small $\varepsilon > 0$, the solution $x_\varepsilon(t)$ is defined on I and $x_\varepsilon(t) \to x(t)$ as $\varepsilon \to 0$ for all $t \in I$. See, for example, Theorem 2.2 of Chapter 2 [41]. Consequently, $x(t) = \lim_{\varepsilon \to 0} x_\varepsilon(t) \leq \bar{x}(t)$. This proves the general case. $\square$

3.3 A More General Existence Result

The method of steps does not work for equations with distributed delay such as

$$x'(t) = -[x(t) - \int_{t-r}^{t} x(s)ds] = -x(t) + \int_{-r}^{0} x(t+\theta)d\theta. \qquad (3.7)$$

or for more exotic equations such as

$$x'(t) = -\max_{t-r \leq s \leq t} x(s) \qquad (3.8)$$

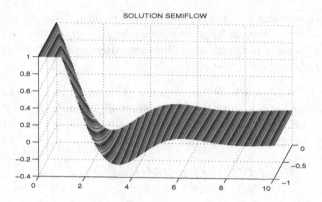

Fig. 3.1 Solution trajectory in C for $x'(t) = -0.75x(t-1)$ with $\phi = \hat{1}$.

A more general approach is necessary. For motivation, let's return to the simpler equation

$$x'(t) = -x(t-r) \tag{3.9}$$

where $r > 0$. If we are given an initial condition:

$$x(t) = \phi(t), -r \leq t \leq 0$$

where $\phi \in C := C([-r,0], \mathbb{R})$, then the method of steps is sufficient to solve the equation for $t \geq 0$. Now this equation is autonomous so we expect it to generate a dynamical system on some state space X: a "solution" should be represented by a curve or trajectory in the state space X. What should this state space be? What should be the state of the system at time t? The state of the system at time $t \geq 0$ should contain all the information necessary to determine the solution for future times $s \geq t$. In particular, it should obviously contain the initial condition ϕ. This argues for $X = C$ as the state space. Apparently, the state of the system at time t cannot be $x(t) \in \mathbb{R}$ because as we know, knowledge of only the value of $x(t)$ is insufficient to determine $x(s)$ for $s \geq t$. For this we must know all the values of $x(\eta)$ for all $\eta \in [t-r, t]$. Equivalently, we must know $x(t+\theta)$, $-r \leq \theta \leq 0$ in order to determine the solution for $s > t$. Therefore, we are led to call the state of the dynamical system at time t the element of C which we write as x_t and define by

$$x_t(\theta) := x(t+\theta), -r \leq \theta \leq 0. \tag{3.10}$$

Thus, we can view the trajectory of our solution as the curve $t \rightarrow x_t$ in the state space C. This turns out to be the "correct" point of view for a dynamical systems framework for these equations. Figure 3.1 depicts the trajectory x_t for $x'(t) = -0.75x(t-1)$. For each $t \in [0, 10]$, the curve depicts x_t. The notation x_t is very convenient for other reasons. For example, we may write Equation (3.9) in the form

$$x'(t) = f(x_t) \tag{3.11}$$

where $f : C \to \mathbb{R}$ is defined by

$$f(\phi) = -\phi(-r)$$

because then, using (3.10), $f(x_t) = -x_t(-r) = -x(t-r)$.

Similarly, equation (3.7) may be written in the form (3.11) where

$$f(\phi) = -\phi(0) + \int_{-r}^{0} \phi(s)ds$$

because then, using (3.10),

$$f(x_t) = -x_t(0) + \int_{-r}^{0} x_t(\theta)d\theta = -x(t) + \int_{-r}^{0} x(t+\theta)d\theta$$

Equation (3.8) may be written in the form (3.11) where

$$f(\phi) = -\max \phi.$$

Our goal should be to obtain the usual existence and uniqueness results for (3.11). More generally, we should consider the initial-value problem for the nonautonomous system

$$x'(t) = f(t, x_t), t \geq \sigma \qquad (3.12)$$
$$x_\sigma = \phi$$

where $\sigma \in \mathbb{R}$ is the initial time and $\phi \in C$ is the state of the system at time σ. This means that

$$x(\sigma + \theta) = \phi(\theta), -r \leq \theta \leq 0$$

or equivalently, that

$$x(t) = \phi(t - \sigma), \sigma - r \leq t \leq \sigma.$$

Furthermore, we should allow x to be a vector in $\mathbb{R}^n$ in (3.12). In this case, the state space should be $C = C([-r,0], \mathbb{R}^n)$, the space of continuous functions from $[-r,0]$ into $\mathbb{R}^n$ and $f : \mathbb{R} \times C \to \mathbb{R}^n$ is a given continuous function. We use $|x|$ for the norm of the vector x and

$$\|\phi\| = \sup\{|\phi(\theta)| : -r \leq \theta \leq 0\}$$

for the norm of $\phi \in C$.

Besides continuity of f, we assume it satisfies a Lipschitz condition on each bounded subset of $\mathbb{R} \times C$.

(Lip) For all $a, b \in \mathbb{R}$ and $M > 0$, there is a $K > 0$ such that:

$$|f(t, \phi) - f(t, \psi)| \leq K \|\phi - \psi\|, a \leq t \leq b, \|\phi\|, \|\psi\| \leq M. \qquad (3.13)$$

Observe that K may depend on the interval a, b and M.

Let's try to find a solution of (3.12) on the interval $[\sigma, \sigma + A]$ for some $A > 0$. Integrating both sides and applying the fundamental theorem of calculus, we get that $x(t)$ should be a continuous solution of the integral equation

$$x(t) = \phi(0) + \int_{\sigma}^{t} f(s, x_s) ds, \sigma \leq t \leq \sigma + A \qquad (3.14)$$

where, of course, $x(t) = \phi(t - \sigma), \sigma - r \leq t \leq \sigma$. We have implicitly used the following (Lemma 2.1, Chapter 2 [41]).

Lemma 3.1. *If* $x : [\sigma - r, \sigma + a] \to \mathbb{R}^n$ *is continuous, then* $t \to x_t$ *is a continuous function from* $[\sigma, \sigma + a]$ *into* $C([-r, 0], \mathbb{R}^n)$.

Proof. As x is uniformly continuous on the closed, bounded interval $I = [\sigma - r, \sigma + a]$, given $\varepsilon > 0$, $\exists \delta > 0$ such that $t, s \in I$, $|t - s| < \delta \Rightarrow |x(t) - x(s)| < \varepsilon$. This implies that

$$|x_t(\theta) - x_s(\theta)| = |x(t + \theta) - x(s + \theta)| < \varepsilon$$

for $\sigma \leq t, s \leq \sigma + a$ with $|t - s| < \delta$ and $-r \leq \theta \leq 0$, proving the result. $\square$

In formulating (3.14), we used that the mapping from $[\sigma, \sigma + A]$ into $\mathbb{R}^n$ given by

$$s \to f(s, x_s)$$

is continuous. This follows because f is continuous and $s \to (s, x_s) \in [\sigma, \sigma + A] \times C$ is continuous by Lemma 3.1.

Our aim is to prove the following result. It is hardly the best possible result; see [41] for sharper results.

Theorem 3.7 *Suppose that* f *is continuous and satisfies the Lipschitz condition* (Lip), $\sigma \in \mathbb{R}$, *and* $M > 0$. *There exists* $A > 0$, *depending only on* M *such that if* $\phi \in C$ *satisfies* $\|\phi\| \leq M$, *then there exists a unique solution* $x(t) = x(t, \phi)$ *of* (3.12), *defined on* $[\sigma - r, \sigma + A]$. *In addition, if* K *is the Lipschitz constant for* f *corresponding to* $[\sigma, \sigma + A]$ *and* M, *then*

$$\max_{\sigma - r \leq \eta \leq \sigma + A} |x(\eta, \phi) - x(\eta, \psi)| \leq \|\phi - \psi\| e^{KA}, \|\phi\|, \|\psi\| \leq M \qquad (3.15)$$

We need the following simple result.

Lemma 3.2. *Let* $f : \mathbb{R} \times C \to \mathbb{R}^n$ *be continuous and satisfy* (Lip). *Then for each finite interval* $[a, b]$ *and* $M > 0$, *there exists* $L > 0$ *such that*

$$|f(t, \psi)| \leq L, t \in [a, b], \|\psi\| \leq M.$$

Proof. If $\hat{0}$ denotes the zero function in C, $\|\psi\| \leq M$, and K is Lipschitz constant for $[a, b]$ and M then

$$|f(t, \psi)| \leq |f(t, \psi) - f(t, \hat{0})| + |f(t, \hat{0})| \leq K\|\psi - \hat{0}\| + |f(t, \hat{0})| \leq KM + P$$

where $P = \max_{a \le s \le b} |f(s, \hat{0})|$. $\square$

Now we turn to the proof of Theorem 3.7.

Proof. Suppose that $\|\phi\| \le M$. Let K be the Lipschitz constant for f on the set $[\sigma, \sigma + r] \times \{\psi \in C : \|\psi\| \le 2M\}$ and let L be the bound on $|f|$ given in Lemma 3.2 for that set. Let $A = \min\{r, M/L\}$.

Given any continuous function $y(t)$ on $[\sigma - r, \sigma + A]$ satisfying $y_\sigma = \phi$ and $|y(t)| \le 2M$ on $[\sigma, \sigma + A]$, we may define a new continuous function z on $[\sigma - r, \sigma + A]$ by

$$z(t) := \phi(0) + \int_\sigma^t f(s, y_s) ds, \sigma \le t \le \sigma + A$$

and $z(t) = \phi(t - \sigma)$ for $t \in [\sigma - r, \sigma]$. It satisfies

$$|z(t)| \le M + L(t - \sigma) \le M + LA \le 2M, \sigma \le t \le \sigma + A$$

Let's use the method of successive approximations to solve (3.14) starting with the initial guess

$$x^{(0)}(t) = \phi(0), \sigma \le t \le \sigma + A$$

and $x^{(0)}(t) = \phi(t - \sigma)$ for $t \in [\sigma - r, \sigma]$. Clearly, $|x^{(0)}(t)| \le M$ on $\sigma \le t \le \sigma + A$.

Now, for $m = 0, 1, 2, \cdots$, define

$$x^{(m+1)}(t) = \phi(0) + \int_\sigma^t f(s, x_s^{(m)}) ds, \sigma \le t \le \sigma + A, \tag{3.16}$$

Again, $x^{(m)}(t) = \phi(t - \sigma)$, $\sigma - r \le t \le \sigma$ so they are defined on $[\sigma - r, \sigma + A]$.

$$|x^{(1)}(t) - x^{(0)}(t)| = |\int_\sigma^t f(s, x_s^{(0)})| ds \le L(t - \sigma), t \in [\sigma, \sigma + A].$$

By (3.16), using (Lip) and $x_\sigma^{(m)} = x_\sigma^{(m-1)}$, we find that

$$|x^{(m+1)}(t) - x^{(m)}(t)| = |\int_\sigma^t [f(s, x_s^{(m)}) - f(s, x_s^{(m-1)})] ds|$$

$$\le K \int_\sigma^t \|x_s^{(m)} - x_s^{(m-1)}\| ds \tag{3.17}$$

$$\le K \int_\sigma^t \sup_{\sigma \le \eta \le s} |x^{(m)}(\eta) - x^{(m-1)}(\eta)| ds.$$

In particular,

$$|x^{(2)}(t) - x^{(1)}(t)| \le K \int_\sigma^t L(s - \sigma) ds = KL(t - \sigma)^2 / 2$$

and

$$|x^{(3)}(t) - x^{(2)}(t)| \le K \int_\sigma^t KL \frac{(s - \sigma)^2}{2} ds = \frac{L}{K} \frac{[K(t - \sigma)]^3}{3!}$$

An induction argument yields that

$$|x^{(m+1)}(t) - x^{(m)}(t)| \le \frac{L}{K} \frac{[K(t-\sigma)]^{m+1}}{(m+1)!}$$

and a triangle inequality argument gives that for $m > n$

$$|x^{(m)}(t) - x^{(n)}(t)| \le |x^{(m)}(t) - x^{(m-1)}(t)| + |x^{(m-1)}(t) - x^{(m-2)}(t)|$$
$$\cdots + |x^{(n+1)}(t) - x^{(n)}(t)|$$
$$\le \frac{L}{K} [\frac{[K(t-\sigma)]^m}{(m)!} + \frac{[K(t-\sigma)]^{m-1}}{(m-1)!}$$
$$\cdots + \frac{L}{K} \frac{[K(t-\sigma)]^{n+1}}{(n+1)!}]$$
$$\le \frac{L}{K} \sum_{j=n+1}^{\infty} \frac{(KA)^j}{j!}$$

As the right-hand side is the tail of a convergent series and so converges to zero as $n \to \infty$, $\{x^{(m)}\}_{m=0}^{\infty}$ is a Cauchy sequence in the space of continuous vector-valued functions on $[\sigma, \sigma + A]$ with the supremum norm. As this space is a complete metric space [65], it follows that there is a continuous function $x : [\sigma, \sigma + A] \to \mathbb{R}^n$ satisfying

$$\sup_{\sigma \le t \le \sigma + A} |x^{(m)}(t) - x(t)| \to 0, m \to \infty$$

Extending $x(t)$ to $[\sigma - r, \sigma + A]$ in the usual way by $x(t) = \phi(t - \sigma)$ for $\sigma - r \le t \le \sigma$, then $x(t)$ satisfies (3.14). To see this first observe that

$$f(s, x_s^{(m)}) \to f(s, x_s), \sigma \le s \le \sigma + A, m \to \infty$$

uniformly because

$$|f(s, x_s^{(m)}) - f(s, x_s)| \le K \|x_s^{(m)} - x_s\| \le K \sup_{\sigma \le t \le \sigma + A} |x^{(m)}(t) - x(t)|$$

This uniform convergence implies that

$$\lim_{m \to \infty} \int_{\sigma}^{t} f(s, x_s^{(m)}) ds = \int_{\sigma}^{t} f(s, x_s) ds.$$

Therefore, on taking limits of both sides of (3.16), we get that (3.14) holds.

Actually, the above proof establishes only the existence of a solution $x(t, \phi)$ on $[\sigma - r, \sigma + A]$ but not uniqueness. If $y : [\sigma - r, \sigma + a]$ is a second solution of the initial-value problem $x_\sigma = \phi$ with $0 < a$, we want to show that y must agree with our solution on $[\sigma, \sigma + \min\{a, A\}]$. To see this, it's first necessary to show that $|y(t)| \le 2M$ on this interval. As $|y(\sigma)| \le M$, if this failed to hold, then there would be a smallest $p < A$ with $|y(p)| = 2M$ so

$$|y(t)| \le |\phi(0)| + \int_\sigma^t |f(s,y_s)|ds \le M + L(t-\sigma), \sigma \le t \le p.$$

Putting $t = p$, we get the contradiction $2M < 2M$ so we conclude that $|y(t)| \le 2M$ on $[\sigma, \sigma + \min\{a, A\}]$. Now we can show that $y(t) = x(t, \phi)$ on this interval by using a Gronwall argument just like the one that gives (3.15).

The assertion (3.15) says that solutions depend continuously on the initial data. It is a consequence of the Gronwall inequality as follows.

$$|x(t,\phi) - x(t,\psi)| \le |\phi(0) - \psi(0)| + |\int_\sigma^t [f(s,x_s(\phi)) - f(s,x_s(\psi))]ds|$$

$$\le \|\phi - \psi\| + K\int_\sigma^t \|x_s(\phi) - x_s(\psi)\|ds$$

$$\le \|\phi - \psi\| + K\int_\sigma^t \max_{\sigma - r \le \eta \le s} |x(\eta,\phi) - x(\eta,\psi)|ds$$

for $\sigma \le t \le \sigma + A$. If we let

$$u(s) := \max_{\sigma - r \le \eta \le s} |x(\eta,\phi) - x(\eta,\psi)|, \sigma \le s \le \sigma + A$$

then we have

$$u(t) \le \|\phi - \psi\| + K\int_\sigma^t u(s)ds, \sigma \le s \le \sigma + A$$

and Gronwall's inequality gives

$$u(t) \le \|\phi - \psi\|e^{K(t-\sigma)}$$

implying that (3.15) holds. Observe that as a special case, we also obtain the inequality

$$\|x_t(\phi) - x_t(\psi)\| \le \|\phi - \psi\|e^{K(t-\sigma)}, \sigma \le t \le \sigma + A. \tag{3.18}$$

$\square$

Remark 3.8 *If f satisfies a global Lipschitz condition, that is, if K in (Lip) can be chosen independent of a, b and M, then we need make no restriction on A in Theorem 3.7. More precisely, its conclusions hold for all $A > 0$. In this case, the solution exists for all $t \ge \sigma$ and (3.18) holds for all $t \ge \sigma$.*

As an example of the application of Theorem 3.7 consider the equation

$$x'(t) = -\alpha x(t) + \beta \tanh(\int_{t-r}^t x(s)ds)$$

which can be viewed as a simple model of a self-excitatory neuron with (distributed) delayed self-excitation. Here, $\alpha, \beta > 0$. The equation may be written in the form (3.11) where

$$f(\phi) = -\alpha\phi(0) + \beta \tanh(\int_{-r}^0 \phi(s)ds)$$

Using the fact that the tanh has a positive derivative of magnitude no larger than one, we find that

$$|f(\phi) - f(\psi)| \leq \alpha|\phi(0) - \psi(0)| + \beta \int_{-r}^{0} |\phi(s) - \psi(s)|ds$$
$$\leq (\alpha + r\beta)\|\phi - \psi\|$$

Therefore f satisfies a global Lipschitz condition (Lip). On the other hand, the delayed logistic equation

$$N'(t) = N(t)[a - b \int_{t-r}^{t} N(s)ds] \tag{3.19}$$

does not satisfy a global Lipschitz condition. Here $a, b > 0$.

Remark 3.9 *Theorem* 3.4 *extends in a natural way to our more general system* (3.12). *We write* $\phi \geq 0$ *for the continuous function* ϕ *provided* $\phi_j(s) \geq 0$, $\forall j, \forall s$. *The key positivity condition is:*

$$\phi \geq 0, \ \phi_i(0) = 0 \Rightarrow f_i(t, \phi) \geq 0 \tag{3.20}$$

See [70, 71] for the proof.

3.4 Continuation of Solutions

Theorem 3.7 provides a local solution of (3.12) defined on $[\sigma - r, \sigma + A]$ for some $A > 0$ whereas in the application we seek a solution defined for all $t \geq \sigma$. We assume (3.13) holds in this section.

Lemma 3.3. *Let* $x : I \to \mathbb{R}^n$ *and* $\hat{x} : J \to \mathbb{R}^n$ *be two solutions of* (3.12), *where* I, J *are intervals of the form* $[\sigma - r, \sigma + \alpha]$ *or* $[\sigma - r, \sigma + \alpha)$ *with* $0 < \alpha \leq \infty$. *Then*

$$x(t) = \hat{x}(t), t \in I \cap J$$

The proof is left to the exercises.

If $x : I \to \mathbb{R}^n$ and $\hat{x} : J \to \mathbb{R}^n$ are two solutions of (3.12) as in Lemma 3.3 and if $I \subset J$, we say that $\hat{x}$ is an *extension* or a *continuation* of x and we write $x \subset \hat{x}$. Let S be the family of all solutions $x : I \to \mathbb{R}^n$ of (3.12) where $I = [\sigma - r, \sigma + \alpha]$ or $I = [\sigma - r, \sigma + \alpha)$ for some $\alpha \in (0, \infty)$. Then $\subset$ is a partial-order relation on S. According to Lemma 3.3, given any two elements $x, \hat{x} \in S$, either $x \subset \hat{x}$ or $\hat{x} \subset x$. A solution $x : I \to \mathbb{R}^n$ is *noncontinuable* if it has no extension to a larger interval. The existence of a unique noncontinuable solution of (3.12) is obtained by a simple argument. Let $J = \cup_{x \in S} \text{domain}(x)$, where $\text{domain}(x)$ refers to the interval on which solution x is defined. Obviously, $J = [\sigma - r, \sigma + \alpha]$ or $J = [\sigma - r, \sigma + \alpha)$ for some $\alpha \in (0, \infty)$ as it is a union of such intervals. But J cannot be the former unless there is $x \in S$ with domain the closed interval $J = [s - r, s + \alpha]$, in which case, we could use Theorem 3.7 to obtain an extension of x to a larger interval, contradicting that x has no extension. Therefore, $J = [\sigma - r, \sigma + \alpha)$ and the unique noncontinuable solution $\bar{x} : J \to \mathbb{R}^n$ of (3.12) can be unambiguously defined by $\bar{x}(t) = x(t)$, where $x \in S$ is any solution defined at t.

If the noncontinuable solution $x : [s - r, s + \alpha) \to \mathbb{R}^n$ of (3.12) satisfies $\alpha < \infty$, then it must "blow up" as $t \nearrow s + \alpha$.

Proposition 3.10 *Assume the hypotheses of Theorem 3.7 hold, $r > 0$, and let $x :$ $[s - r, s + \alpha) \to \mathbb{R}^n$, where $0 < \alpha \leq \infty$, be the unique noncontinuable solution of (3.12). If $\alpha < \infty$, then*

$$\lim_{t \nearrow s + \alpha} \|x_t\| = \infty$$

Proof. If not $\exists M > 0$ and an increasing sequence $\{t_m\}_m$ satisfying $t_m < \alpha$ and $t_m \to \alpha$ such that $\|x_{s+t_m}\| \leq M$. This means that for each m, $|x(t)| \leq M$, $t \in I_m = [s + t_m - r, s + t_m]$. But $t_{m'} - t_m < r$ when $m, m' \geq K$ for some K so I_m and $I_{m'}$ overlap and $\cup_{m \geq K} I_m = [t_K + s - r, s + \alpha)$. Therefore, there exists $N > 0$ such that $|x(t)| \leq N$, $t \in [s - r, s + \alpha)$ and, in particular, $\|x_t\| \leq N, s \leq t < s + \alpha$. It follows from Lemma 3.2 that there exists $P > 0$ such that $|x'(t)| = |f(t, x_t)| \leq P$, $s \leq t < a + \alpha$. This implies that $x : [s, s + \alpha) \to \mathbb{R}^n$ is uniformly continuous because $|x(t) - x(t')| = |\int_t^{t'} x'(s) ds| \leq P|t - t'|$ for $t, t' \in [s, s + \alpha)$, $t < t'$. Consequently, the limit $X = \lim_{t \to s + \alpha} x(t)$ exists and $x : [s - r, s + \alpha] \to \mathbb{R}^n$, defined at $t = s + \alpha$ to be X, is continuous. Because $x(t)$ satisfies the integral equation (3.14) for $t \in [s, s + \alpha)$ and because it extends continuously to the interval $[s - r, s + \alpha]$, we can take limits in (3.14) as $t \to s + \alpha$ to see that (3.14) holds for the closed interval $[s - r, s + \alpha]$. This means that $x : [s - r, s + \alpha] \to \mathbb{R}^n$ is a solution of (3.12). But this contradicts that $x : [s - r, s + \alpha) \to \mathbb{R}^n$ is noncontinuable. $\square$

3.5 Remarks on Backward Continuation

Backward continuation refers to solving (3.12) for $t < \sigma$. Recall that up until now all our efforts were devoted to solving the initial-value problem (3.12) for $t > \sigma$. A glance at some of the delay differential equations we have introduced in Chapter 1 should convince the reader of the fundamental asymmetry between the future and the past for delay equations, an asymmetry that is absent from ODEs. For ODEs, we do not make any fundamental distinction between solving forward in time or backward in time.

Following [41], if $\alpha > 0$ then we say that $x : [\sigma - r - \alpha, \sigma] \to \mathbb{R}^n$ is a (backward) solution of the initial-value problem (3.12) if x is continuous, $x_\sigma = \phi$, and x satisfies

$$x'(t) = f(t, x_t), t \in [\sigma - \alpha, \sigma]$$

Because $x_\sigma = \phi$ must hold as well as the equation above, we see that $x(t) = \phi(t - \sigma)$ must be continuously differentiable for $t \in [\sigma - \alpha, \sigma]$, or equivalently ϕ must be continuously differentiable on $[-\alpha, 0]$ (or on all $[-r, 0]$ if $\alpha \geq r$). Obviously, this means ϕ must be quite special among elements of C (the typical element of C will not be differentiable.). Even more, because the equation above holds at $t = \sigma$ we must have

$$\phi'(0) = x'(\sigma) = f(\sigma, x_\sigma) = f(\sigma, \phi) \qquad (3.21)$$

where $\phi'(0)$ denotes the left-hand derivative at 0. This compatibility condition is reminiscent of the compatibility conditions that arise for boundary-value problems in PDEs. In any case, this means that ϕ is really really special; it must belong to the presumably "thin" submanifold M of C where:

$$M = \{\psi \in C : \psi'(0) = f(\sigma, \psi)\}.$$

We conclude from these remarks that for the typical initial function $\phi \in C$ there will not exist a backward continuation of the initial value problem (3.12). Nevertheless, Theorem 5.1 of [41] gives (very technical) sufficient conditions for the existence of backward solutions for these very special initial data.

We should keep in mind, however, that if $x : [\sigma - r, \sigma + A)$, $A > 0$ is a (forward) solution of (3.12) and if $\sigma_1 \in (\sigma, \sigma + A)$ then the initial-value problem corresponding to the initial condition (σ_1, ψ) where $\psi = x_{\sigma_1}$ has a backward solution, namely x. Later, we show that many solutions of autonomous systems extend to all $t \in \mathbb{R}$. Steady-state and periodic solutions are specific examples.

3.6 Stability Definitions

Our definitions follow ones in Hale and Lunel [41]. We consider the system of delay differential equations

$$x'(t) = f(t, x_t)$$

Suppose that it satisfies $f(t, 0) = 0$, $t \in \mathbb{R}$ so that $x(t) = 0$ is a solution. The solution $x = 0$ is *stable* if for any $\sigma \in \mathbb{R}$ and $\varepsilon > 0$, there exists $\delta = \delta(\sigma, \varepsilon) > 0$ such that $\phi \in C$ and $\|\phi\| < \delta$ implies that $\|x_t(\sigma, \phi)\| < \varepsilon$, $t \geq \sigma$ where $x(t, \sigma, \phi)$ is our notation for the solution of (3.12). It is *asymptotically stable* if it is stable and if there exists $b(\sigma) > 0$ such that whenever $\phi \in C$ and $\|\phi\| < b(\sigma)$, then $x(t, \sigma, \phi) \to 0$, $t \to \infty$. Finally, $x = 0$ is *unstable* if it is not stable.

The stability of any other solution of (3.12) can be defined by changing variables such that the given solution is the zero solution. More precisely, given a solution $y(t)$ of (3.12) defined on $t \in \mathbb{R}$, its stability properties are those of the zero solution of

$$z'(t) = f(t, z_t + y_t) - f(t, y_t) \qquad (3.22)$$

Indeed, if $x(t)$ is another solution of (3.12) and if we let $z(t) = x(t) - y(t)$ then $z_t = x_t - y_t$ so z satisfies (3.22).

The special case that solution $y(t) \equiv e$, an equilibrium, is of primary interest. In that case, let $\hat{e} \in C$ be the constant function identically equal to e. Then equation

(3.22) for the perturbation $z(t) = x(t) - e$ becomes

$$z'(t) = f(t, z_t + \hat{e})$$

Note that by the change of variables the equilibrium $y(t) = e$ now becomes $z(t) = 0$.

Exercises

Exercise 3.1. Prove Theorem 3.1. Hint: Use standard ODE results.

Exercise 3.2. Which of the systems among (1.3), (1.8), (1.9), (1.16), (1.18), (1.17), (8), (1.10), have the property that nonnegative initial data always give rise to non-negative solutions? If an equation fails to have this property, provide nonnegative initial data ϕ and s such that the corresponding solution has a negative component for some values of t.

Exercise 3.3. Aiello and Freedman [2] introduce a model of a stage-structured population consisting of immature x_1 and mature x_2 individuals:

$$x_1'(t) = rx_2(t) - dx_1(t) - \beta e^{-d\tau} x_2(t - \tau)$$
$$x_2'(t) = \beta e^{-d\tau} x_2(t - \tau) - ax_2^2(t)$$

Do nonnegative initial data give rise to nonnegative solutions?

Exercise 3.4. Combine Theorems 3.2 and 3.4 to show that every solution of (1.3) is defined for all $t \geq s$ provided the initial data ϕ are nonnegative. Hint: The delayed recruitment rate is bounded. Use differential inequalities or Gronwall's lemma.

Exercise 3.5. Find an explicit formula for $f(t, x_t)$ for the glucose-insulin system (3.23).

$$G'(t) = -aG(t) - bI(t)G(t) + c(t)$$
$$I'(t) = -dI(t) + e \int_{t-\tau}^{t} G(s)ds$$

Exercise 3.6. Is condition (Lip) satisfied in the examples (3.7)-(3.9)? Find K.

Exercise 3.7. Let $x : [s-r, s+A] \to \mathbb{R}^n$ be the solution of (3.12) guaranteed by Theorem 3.7. Show that there is $P > 0$ such that $|x(t)| \leq P$, $t \in [s-r, s+A]$.

Exercise 3.8. Show that Theorem 3.7 applies to the initial value problem for (3.19).

Exercise 3.9. Use Remark 3.9 to show that solutions of the Equations (1.2) and (3.23) corresponding to nonnegative initial data (in all components) are nonnegative in the future.

Exercise 3.10. Prove Lemma 3.3. Hint: Use Gronwall's inequality and (3.13).

Chapter 4
Linear Systems and Linearization

Abstract Key to the analysis of nonlinear systems is determining the stability of the equilibria. The classical method of determining stability is to linearize the system about the equilibrium and to determine exponential rates of growth and decay for the associated linear system. The framework for carrying this out is taken up in this chapter. Although the method is similar to that for ODEs, the characteristic equation is more complicated, typically having infinitely many roots. Fortunately, all but finitely many of these roots have real part less than any given real number.

4.1 Autonomous Linear Systems

Even when one is only interested in real-valued solutions of systems, it is useful to allow for complex-valued solutions. Therefore, in this chapter, we allow for this by modifying our space C as $C = C([-r,0], \mathbb{C}^n)$. A function $L : C \to \mathbb{C}^n$ is linear if it satisfies

$$L(a\phi + b\psi) = aL(\phi) + bL(\psi), \phi, \psi \in C, a, b \in \mathbb{C}.$$

L is said to be *bounded* if there exists $K > 0$ such that

$$|L(\phi)| \leq K\|\phi\|, \phi \in C$$

Our aim in this chapter is to consider some aspects of the linear delay differential equation

$$x'(t) = L(x_t) \tag{4.1}$$

We assume without further mention that L is bounded and linear.

An important example is the discrete-delay case. Let A and B be $n \times n$ matrices and define

$$L(\phi) = A\phi(0) + B\phi(-r).$$

Then

$$|L(\phi)| \leq |A||\phi(0)| + |B||\phi(-r)| \leq (|A| + |B|)\|\phi\|$$

H. Smith, *An Introduction to Delay Differential Equations with Applications to the Life Sciences*, Texts in Applied Mathematics 57, DOI 10.1007/978-1-4419-7646-8_4,
© Springer Science+Business Media, LLC 2011

and consequently L is bounded. Equation (4.1) takes the form

$$x'(t) = Ax(t) + Bx(t-r) \tag{4.2}$$

In the case of more than one delay, we have the linear system

$$x'(t) = Ax(t) + \sum_{j=1}^{m} B_j x(t-r_j) \tag{4.3}$$

where A, B_j are matrices and $r_j \geq 0$.

Equation (4.1) is clearly an autonomous system so we may as well restrict initial data to prescribing the values of x on $[-r, 0]$:

$$x(t) = \phi(t), -r \leq t \leq 0 \tag{4.4}$$

where $\phi \in C$.

Because L is bounded, it satisfies a global Lipschitz condition

$$|L(\phi) - L(\psi)| = |L(\phi - \psi)| \leq K\|\phi - \psi\|$$

Consequently, Theorem 3.7 and Remark 3.8 apply to the initial-value problem (4.1) and (4.4). Therefore, there exists a unique maximally defined solution $x : [-r, \infty) \to \mathbb{C}^n$ defined for all $t \geq 0$.

The linear character of (4.1) implies the usual *superposition principle*: a linear combination of solutions is again a solution.

Proposition 4.1 *Let $x(t, \phi)$ denote the solution of (4.1),(4.4). Then the map $\phi \to x(t, \phi)$ is linear:*

$$x(t, a\phi + b\psi) = ax(t, \phi) + bx(t, \psi), t \geq 0, \phi, \psi \in C, a, b \in \mathbb{C}.$$

Our formulation of (4.1) includes equations with distributed delays as well. Let $r_{i,j}$, $i, j = 1, 2$ be positive and let $k_{ij} : [0, r_{ij}] \to \mathbb{C}$ be integrable functions. Consider the linear system

$$x_1'(t) = \int_0^{r_{11}} k_{11}(s)x_1(t-s)ds + \int_0^{r_{12}} k_{12}(s)x_2(t-s)ds$$
$$x_2'(t) = \int_0^{r_{21}} k_{21}(s)x_1(t-s)ds + \int_0^{r_{22}} k_{22}(s)x_2(t-s)ds \tag{4.5}$$

Let $r = \max r_{ij}$ and extend k_{ij} to $[0, r]$ if necessary by making it identically zero on $(r_{ij}, r]$. Define the matrix-valued function k on $[0, r]$ so $k(s) = (k_{ij}(s))$ and let $x(t) = (x_1(t), x_2(t))$. Then we may write (4.5) as

$$x'(t) = \int_0^r k(s)x(t-s)ds \tag{4.6}$$

4.2 Laplace Transform and Variation of Constants Formula

We specialize to the case of the nonhomogeneous initial value problem

$$x'(t) = Ax(t) + Bx(t-r) + f(t), t \geq 0, x_0 = \phi \qquad (4.7)$$

where A, B are scalars; the case that they are matrices is treated by analogy. Our goal is to obtain a variation-of-constants formula for the nonhomogeneous equation and we follow [41] in using the Laplace transform:

$$F(s) = \mathcal{L}f = \int_0^\infty e^{-st} f(t) dt$$

where $f : [0, \infty) \to \mathbb{C}$ is an exponentially bounded function, that is, $|f(t)| \leq Me^{kt}$ for some real M, k. We follow the usual notation that the upper case F is the Laplace transform of lower case f. The key property of the Laplace transform that we exploit is that the transform of a convolution is the product of the transforms. If $f, g : [0, \infty) \to \mathbb{C}$ then their convolution is defined by

$$(f * g)(t) = \int_0^t f(\tau)g(t-\tau)d\tau = \int_0^t f(t-\tau)g(\tau)d\tau$$

The Laplace transform satisfies

$$\mathcal{L}(f * g) = F(s)G(s)$$

where $F = \mathcal{L}f$ and $G = \mathcal{L}g$.

Applying the transform to (4.7), we obtain

$$sX(s) - \phi(0) = AX(s) + B[\int_0^r e^{-st}\phi(t-r)dt$$
$$+ \int_r^\infty e^{-st}x(t-r)dt] + F(s)$$
$$= AX(s) + B[\int_0^r e^{-st}\phi(t-r)dt + e^{-sr}X(s)] + F(s)$$
$$= [A + e^{-sr}B]X + B\int_0^r e^{-st}\phi(t-r)dt + F(s)$$
$$= [A + e^{-sr}B]X + B\Phi(s) + F(s)$$

where $\Phi = \mathcal{L}(\phi(\cdot - r))$ and where we have extended ϕ to $[-r, \infty)$ by making it zero for $t > 0$. This leads to

$$X(s) = K(s)[\phi(0) + B\Phi(s) + F(s)] \qquad (4.8)$$

where

$$K(s) = (s - A - e^{-sr}B)^{-1}$$

In order to make use of the convolution result, we need to know the inverse transform k of K. In view of the calculations above, we see that k is the solution of (4.7), with $f = 0$, for the initial data

$$\xi(\theta) = \left\{ \begin{matrix} 1, & \theta = 0 \\ 0, & -r \le \theta < 0 \end{matrix} \right\} \tag{4.9}$$

In spite of the discontinuity of ξ at zero, the method of steps readily establishes that the solution k exists for $t \ge 0$. In fact,

$$k(t) = e^{At}, 0 \le t < r$$

and it satisfies the initial value problem:

$$x'(t) = Ax(t) + Be^{A(t-r)}, x(r) = e^{Ar}, r \le t < 2r$$

It is not hard to see that k is continuous for $t \ge 0$, continuously differentiable for $t \ge r$, twice continuously differentiable for $t \ge 2r$, and so on. Solution k is called the fundamental solution of (4.7).

By Exercise 4.2, we may express the solution of (4.7) as

$$x(t) = x(t; \phi, f) = x(t; \phi, 0) + x(t; 0, f)$$

where the first summand satisfies (4.2) and $x_0 = \phi$. In view of our calculations above and the convolution law of the Laplace transform, we have

$$x(t; 0, f) = \int_0^t k(t - \tau) f(\tau) d\tau \tag{4.10}$$

It can be verified directly that (4.10) agrees with $x(t; 0, f)$.

Similar arguments lead to

$$x(t; \phi, 0) = k(t)\phi(0) + \int_0^t k(t - \tau) B\phi(\tau - r) d\tau \tag{4.11}$$

These formulas hold with minor changes in the case that (4.7) is a vector system with matrices A, B. In this case,

$$K(s) = (sI - A - e^{-sr}B)^{-1}$$

is a matrix-valued function, I being the identity matrix, and therefore K is too. The initial data (4.9) are matrix-valued as well with 1 replaced by the identity matrix I. Matrix-valued solution k is the fundamental matrix solution in this case.

4.3 The Characteristic Equation

We seek exponentially growing solutions of (4.1) of the form

$$x(t) = e^{\lambda t}v, v \neq 0$$

where λ is complex and v is a vector whose components are complex. It is useful to have the notation $\exp_\lambda$ for the continuous function defined on $[-r, 0]$ by $\exp_\lambda(\theta) = e^{\lambda\theta}$. Using it, we see that the state x_t corresponding to $x(t)$ is $x_t = e^{\lambda t}(\exp_\lambda)v$ because

$$x_t(\theta) = x(t + \theta) = e^{\lambda(t+\theta)}v = e^{\lambda t}\exp_\lambda(\theta)v$$

For $x(t)$ to be a solution, we must have

$$x'(t) = \lambda e^{\lambda t}v = L(x_t) = e^{\lambda t}L(\exp_\lambda v)$$

or

$$\lambda v = L(\exp_\lambda v)$$

Writing $v = \sum_j v_j e_j$ where $\{e_j\}_j$ is the standard basis for $\mathbb{C}^n$, then $L(\exp_\lambda v) = \sum_j v_j L(\exp_\lambda e_j)$. Define L_λ to be the $n \times n$ matrix

$$L_\lambda = (L(\exp_\lambda e_1)|L(\exp_\lambda e_2)|\cdots|L(\exp_\lambda e_n)) = (L_i(\exp_\lambda e_j))$$

where $L_i(\phi)$ is component i of $L(\phi)$. Then $L(\exp_\lambda v) = L_\lambda v$ and we see that $x(t) = e^{\lambda t}v$ is a nonzero solution of (4.1) if λ is a solution of the *characteristic equation*:

$$\det(\lambda I - L_\lambda) = 0 \tag{4.12}$$

In this case, $v \neq 0$ must belong to the null space of $\lambda I - L_\lambda$. We sometimes refer to a solution $\lambda \in \mathbb{C}$ of (4.12) as a *characteristic root*.

For the special case (4.2), the characteristic equation is

$$\det[\lambda I - A - e^{-\lambda r}B] = 0 \tag{4.13}$$

If A and B are 2×2 matrices, (4.13) may be expressed as $P(\lambda, e^{-r\lambda}) = 0$ where $P(z, w)$ is a quadratic polynomial in (z, w):

$$\lambda^2 - (tr(A))\lambda + \det(A) + e^{-2r\lambda}\det(B) + e^{-r\lambda}[C - \lambda(tr(B))] = P(z, w) \tag{4.14}$$

and

$$C = \det(a^1|b^2) + \det(b^1|a^2)$$

Here, $\det(A)$ denotes the determinant of A, $tr(A)$ denotes its trace, and $(a^1|b^2)$ denotes the matrix with the first column from A and second column from B. It is not hard to see that (4.13) may be expressed as $P(\lambda, e^{-r\lambda}) = 0$ where $P(z, w)$ is a polynomial of degree n in (z, w) in the general case $n \geq 2$.

The first observation is that h, defined by $h(\lambda) = \det(\lambda I - L_\lambda)$, is an analytic function defined for all $\lambda \in \mathbb{C}$, that is, an entire function, analytic in the entire complex plane. See [1, 17] and Appendix A for definitions.

Lemma 4.1. $h(\lambda) = \det(\lambda I - L_\lambda)$ *is an entire function.*

Proof. The determinant of a matrix can be expressed as a sum of products of entries of the matrix. Because sums and products of analytic functions are analytic, it follows that the determinant of a matrix whose entries are analytic is itself analytic. It suffices to show that for each complex vector v, the function g defined by $g(\lambda) = L(\exp_\lambda v) \in \mathbb{C}^n$ is differentiable with respect to λ. Consider the difference quotient

$$D(z) = \frac{g(\lambda + z) - g(\lambda)}{z}$$

Using the linearity of L, it may be written as

$$D(z) = L(\exp_\lambda k(z)v)$$

where $k(z)(\theta) = (e^{z\theta} - 1)/z$. Because $k(z) \to q$ as $z \to 0$, uniformly on $[-r, 0]$, where $q(\theta) = \theta$, it follows that $D(z) \to L(\exp_\lambda qv)$ as $z \to 0$. □

Properties of nontrivial entire functions, in particular of h, are listed below:

 (i) Each characteristic root has finite order.
 (ii) There are at most countably many characteristic roots.
(iii) The set of characteristic roots has no finite accumulation point.

Remarkably, there are only finitely many characteristic roots with positive real part.

Lemma 4.2. *Given $\sigma \in \mathbb{R}$, there are at most finitely many characteristic roots satisfying $\Re(\lambda) > \sigma$.*

If there are infinitely many distinct characteristic roots $\{\lambda_n\}_n$, then

$$\Re(\lambda_n) \to -\infty, n \to \infty$$

Proof. If false, then there is infinite sequence $\{\lambda_j\}$ of characteristic roots with $\Re(\lambda_j) > \sigma$ and, by (iii) and the Bolzano–Weierstrass theorem, $|\lambda_j| \to \infty$ as $j \to \infty$. This means that $C_j = I - (\lambda_j)^{-1} L_{\lambda_j}$ is a singular matrix. But $C_j \to I$ as $j \to \infty$ because

$$|L(\exp_{\lambda_j} v)| \le K \|\exp_{\lambda_j} v\|$$
$$\le K|v| \max\{|e^{\lambda_j \theta}| : -r \le \theta \le 0\}$$
$$\le K|v| \max\{e^{\sigma\theta} : -r \le \theta \le 0\}$$

is bounded independently of j, implying that $(\lambda_j)^{-1} L_{\lambda_j} \to 0$. By continuity of the determinant, $0 = \det(C_j) \to \det(I) = 1$ as $j \to \infty$, a contradiction. □

An important implication of Lemma 4.2 is that there exists $\sigma \in \mathbb{R}$ and a finite set of "dominant characteristic roots" having maximal real part equal to σ with all other roots having real part strictly less than σ.

In the applications, complex characteristic roots come in conjugate pairs.

Proposition 4.2 *Suppose that L maps real functions to real vectors:* $L(C([-r,0],\mathbb{R}^n)) \subset \mathbb{R}^n$. *Then* λ *is a characteristic root if and only if* $\bar{\lambda}$ *is a characteristic root.*

Proof. Our hypotheses imply that $\overline{L(\phi)} = L(\bar{\phi})$ inasmuch as

$$\overline{L(u+iv)} = \overline{Lu+iLv} = Lu - iLv = L(u-iv) = L(\overline{u+iv}).$$

Because $\overline{e^\lambda} = e^{\bar{\lambda}}$ we have $\overline{L(\exp_\lambda e_j)} = L(\overline{\exp_\lambda e_j}) = L(\exp_{\bar{\lambda}} e_j)$, implying that $\overline{L_\lambda} = L_{\bar{\lambda}}$. Therefore, if λ is a root, then

$$\begin{aligned}
0 &= \overline{\det(\lambda I - L_\lambda)} \\
&= \det(\overline{\lambda I - L_\lambda}) \\
&= \det(\bar{\lambda} I - L_{\bar{\lambda}})
\end{aligned}$$

so $\bar{\lambda}$ is a root. $\square$

The main result of this section concerns the stability of the $x = 0$ solution of (4.1).

Theorem 4.3 *Suppose that* $\Re(\lambda) < \mu$ *for every characteristic root* λ. *Then there exists* $K > 0$ *such that*

$$|x(t,\phi)| \le K e^{\mu t} \|\phi\|, t \ge 0, \phi \in C \tag{4.15}$$

where $x(t,\phi)$ *is the solution of* (4.1) *satisfying* $x_0 = \phi$.

In particular, $x = 0$ *is asymptotically stable for* (4.1) *if* $\Re(\lambda) < 0$ *for every characteristic root; it is unstable if there is a root satisfying* $\Re(\lambda) > 0$.

Proof. We sketch the proof of estimate (4.15) for the special case of system (4.2) using the representation (4.11):

$$x(t) = x(t;\phi,0) = k(t)\phi(0) + \int_0^t k(t-\tau)B\phi(\tau-r)d\tau$$

It is enough to show that there exists μ, K as above such that

$$|k(t)| \le K e^{\mu t}, t \ge 0 \tag{4.16}$$

because this implies that

$$\begin{aligned}
|x(t)| &= K e^{\mu t} |\phi(0)| + K e^{\mu t} |B| \|\phi\| \int_0^r e^{-\mu\tau} d\tau \\
&= K e^{\mu t} \|\phi\| \left[1 + |B| \int_0^r e^{-\mu\tau} d\tau \right]
\end{aligned}$$

Estimate (4.16) requires use of the contour integral representation of $\mathcal{L}^{-1}$ and careful estimates. See Chapter 1, Theorem 5.2 in [41].   □

4.4 Small Delays Are Harmless

Consider the linear delay system

$$z'(t) = Az(t) + Bz(t - r)$$

and its nondelayed counterpart obtained by setting $r = 0$,

$$z'(t) = (A + B)z(t)$$

We want to explore the correspondence of the characteristic roots of

$$h(\lambda, r) = \det[\lambda I - A - e^{-\lambda r}B] = 0 \tag{4.17}$$

to the eigenvalues of $A + B$:

$$h(\lambda, 0) = \det[\lambda I - A - B] = 0$$

Theorem 4.4 *Let $z_1, z_2, \ldots, z_k$ be the distinct eigenvalues of $A + B$, let $\delta > 0$, and let $s \in \mathbb{R}$ satisfy $s < \min_i \Re(z_i)$. Then there exists $r_0 > 0$ such that if $0 < r < r_0$ and $h(z, r) = 0$ for some z then either $\Re(z) < s$ or $|z - z_i| < \delta$ for some i.*

In words, for small enough delay, the characteristic roots of (4.17) are either very near the eigenvalues of $A + B$ or have more negative real parts than any of the eigenvalues of $A + B$. We may choose s as negative as we like, therefore we can force these latter roots to have very negative real parts if r is small enough.

Proof. Let $G = \{z \in \mathbb{C} : \Re(z) \geq s, |z - z_i| \geq \delta, 1 \leq i \leq k\}$. Our goal is to show that there are no characteristic roots in the closed set G if r is small enough. If this were false, then there exists a sequence $\{r_n\}_n$ of delays with $r_n > 0$ and $r_n \to 0$ and a corresponding sequence of characteristic roots $z_n \in G$ so that $h(z_n, r_n) = 0$. By the Bolzano–Weierstrass theorem [65] there are two cases: (1) $\{z_n\}$ has a convergent subsequence converging to $\bar{z} \in G$ (G is closed), or (2) $|z_n| \to \infty$. In case (1) continuity of h implies that $h(\bar{z}, 0) = 0$ but $A + B$ has no eigenvalues in G. This case cannot occur. We conclude that case (2) must hold. Now we argue exactly as in the proof of Lemma 4.2 that $C_n = I - (1/z_n)[A + e^{-z_n r_n}B] \to I$ because $\Re(z_n) \geq s$ and $|z_n| \to \infty$, so it cannot be singular for large n. This contradiction proves the result.   □

Remark 4.5 *Theorem 6.8 continues to hold if $A = A(r)$ and $B = B(r)$ depend continuously on r and the z_j are the eigenvalues of $A(0) + B(0)$.*

What about those characteristic roots that do get very near to one of the z_i as $r \to 0$? They can be described by the implicit function theorem A.3 if z_i is an order-one root.

Theorem 6.8 extends in the obvious way to (4.3).

Small delays are harmless in the sense that if asymptotic stability holds when $\tau = 0$, then it continues to hold for small delays inasmuch as we may choose δ small enough that the δ-ball about each eigenvalue of $A + B$ belongs to the left half-plane and we may choose s negative. On the other hand, if instability holds for $\tau = 0$ due to a simple positive root or a complex conjugate pair of roots with positive real part, then the implicit function theorem may be applied to show that instability continues to hold for small $r > 0$.

The recent monograph [59] describes computational methods for determining the critical (largest real-part) characteristic roots.

4.5 The Scalar Equation $x'(t) = Ax(t) + Bx(t - r)$

In this section, we obtain a more complete picture of the characteristic roots associated with (4.2) in the case that A and B are real scalars. This corresponds to the equation

$$x'(t) = Ax(t) + Bx(t - r) \tag{4.18}$$

In that case, (4.13) becomes

$$\lambda = A + Be^{-\lambda r} \tag{4.19}$$

Multiply by r and let

$$z = r\lambda, \alpha = Ar, \beta = Br$$

to obtain:

$$z = \alpha + \beta e^{-z}. \tag{4.20}$$

If we write $z = x + iy$, then the equations are

$$0 = x - \alpha - \beta e^{-x}\cos(y) \tag{4.21}$$
$$0 = y + \beta e^{-x}\sin(y)$$

Observe that $z = 0$ is a root precisely when $\alpha + \beta = 0$ and a portion of this line is plotted in Figure 4.1 below.

Define the function $F(z, \alpha, \beta) := z - \alpha - \beta e^{-z}$ whose zeros are the roots of (4.20). In Exercise 4.7 we identify the branch of solutions of (4.20) on which the implicit function theorem fails. At all other solutions (z_0, α_0, β_0) of (4.20), the implicit function theorem A.3 guarantees a smooth root $z = z(\alpha, \beta)$ for (α, β) near (α_0, β_0).

Setting $x = 0$ and solving for α and β gives the "neutral stability curves" in parameter space along which (4.20) has purely imaginary roots $z = iy$.

$$\alpha = y\cos(y)/\sin(y) \tag{4.22}$$
$$\beta = -y/\sin(y)$$

As roots come in complex conjugate pairs, we may restrict $y \geq 0$. Notice that the curve is welldefined at $y = 0$ where $(\alpha, \beta) = (1, -1)$, which coincides with the parameter values at which $z = 0$ is a double root. We denote by

$$C_0 := \{(\alpha, \beta) = (y\cos(y)/\sin(y), -y/\sin(y)), \ 0 \leq y < \pi\}$$

the curve along which $z = \pm iy$, $0 \leq y < \pi$ are roots. It is depicted in Figure 4.1 below. Easy calculations give

$$\frac{d\alpha}{dy} < 0, \frac{d\beta}{dy} < 0, 0 < y < \pi$$

so both $\alpha(y), \beta(y)$ decrease with increasing y. Starting from $(1, -1)$ when $y = 0$, $(\alpha(y), \beta(y))$ meets the β-axis at $(0, -\pi/2)$ when $y = \pi/2$. It then enters the third quadrant and approaches $(-\infty, -\infty)$ from below and tangent to the line $\alpha = \beta$ as $y \nearrow \pi$ because $\alpha/\beta = -\cos(y) \to 1$ whereas both $\alpha, \beta \to -\infty$ as $y \nearrow \pi$.

We also need to consider the curves

$$C_n := \{(\alpha, \beta) = (y\cos(y)/\sin(y), -y/\sin(y)), \ n\pi < y < (n+1)\pi\}, n \geq 1,$$

where $z = \pm iy$, $n\pi < y < (n+1)\pi$ are roots. Notice that $(-1)^n \sin(y) > 0$ on $n\pi < y < (n+1)\pi$ so $(-1)^{n+1}\beta > 0$ on C_n but α changes sign at $y = n\pi + \pi/2$. On C_n, $d\alpha/dy < 0$ but $d\beta/dy$ changes sign on $(n\pi, (n+1)\pi)$ where $\tan(y) = y$. Because $\beta/\alpha = -1/\cos(y)$, $|\beta/\alpha| > 1$ on C_n implying that $C_1, C_3, \ldots$ lie strictly above the graph of $\beta = |\alpha|$ and $C_2, C_4, \ldots$ lie strictly below the graph of $\beta = -|\alpha|$. It is easy to see that C_{2n+1} lies strictly above C_{2n-1} for $n = 1, 2, \ldots$ and that $C_{2(n+1)}$ lies strictly below C_{2n} for $n = 1, 2, \ldots$. See Figure XI.1, page 306, of Diekmann et al. [26]. C_n never meets C_0, the line $\alpha + \beta = 0$, nor the open region enclosed by these two curves.

Let

$$R(\alpha, \beta) = \{z \in \mathbb{C} : \Re(z) > 0, \ F(z, \alpha, \beta) = 0\}$$

be the set of "unstable roots" for given parameter pair (α, β) and let

$$I = \{(\alpha, \beta) : F(iy, \alpha, \beta) = 0 \text{ for some real } y\}$$
$$= \{(\alpha, \beta) : \alpha + \beta = 0\} \cup (\cup_{n \geq 0} C_n)$$

be the parameter set where purely imaginary or zero roots exist. Finally, define

$$Z(\alpha, \beta) = \sum_{z \in R(\alpha, \beta)} \text{order of root } z$$

which counts the "unstable roots" according to multiplicity. This is a finite sum according to Lemma 4.2.

We show that Z is continuous except possibly at points (α, β) where $F(z, \alpha, \beta) = 0$ has purely imaginary or zero roots:

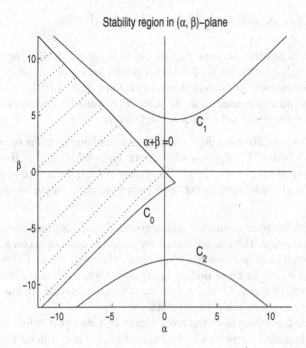

Fig. 4.1 Stability region for (4.20) in the (α, β)-plane lies to the left of the displayed curve.

Lemma 4.3. *The integer-valued function $Z(\alpha, \beta)$ is continuous at all points (α, β) that do not belong to the closed set I. Consequently, Z is constant on the connected components of the complement of I.*

Proof. As a first step, we give bounds for any root $z = x + iy$ with $x = \Re(z) \geq 0$ in terms of α and β. From (4.21) we see that such a root must satisfy

$$0 \leq x \leq |\alpha| + |\beta|, |y| \leq |\beta|.$$

Now fix (α_0, β_0) not in I. As the complement of I is open, we can find a closed ball B_0 centered at (α_0, β_0) so small that B_0 does not intersect I. By the estimate above, we can find $M > 0$ such that any root $z = x + iy$ with $x > 0$ corresponding to any $(\alpha, \beta) \in B_0$ satisfies $x, |y| \leq M$. Thus, any such root lies inside the simple closed curve γ, oriented counterclockwise, bounding the rectangle $R := [0, 2M] \times [-2M, 2M]$ in the complex plane. Hence, for $(\alpha, \beta) \in B_0$

$$Z(\alpha, \beta) = \sum_{z \text{ inside } \gamma} \text{order of root } z$$

Furthermore, there are no roots of $F(z, \alpha_0, \beta_0) = 0$ on γ itself because (α_0, β_0) does not belong to I. By shrinking the radius of the closed ball B_0, if necessary, we can assume that for z on γ:

$$|F(z,\alpha,\beta) - F(z,\alpha_0,\beta_0)| < \min_{z \in \gamma}\{|F(z,\alpha_0,\beta_0)|\}, (\alpha,\beta) \in B_0.$$

Note that the minimum is positive because $F(z,\alpha_0,\beta_0) \neq 0$ for $z \in \gamma$. By Theorem A.4, $Z(\alpha,\beta) = Z(\alpha_0,\beta_0)$ for all $(\alpha,\beta) \in B_0$. This proves the continuity of Z.

Let J be a connected component of the complement of I. Then $Z : J \rightarrow \{0,1,2,\dots\}$ is a continuous function and J is connected so its image $Z(J)$ must be a connected subset of the nonnegative integers. So it must be a single integer. $\square$

Proposition 4.6 *All roots of* (4.20) *have* $\Re(z) < 0$ *for* (α,β) *belonging to the open region bounded below by curve* C_0 *and bounded above by curve* $\{(\alpha,\beta) : \beta = -\alpha, \ \alpha \leq 1\}$ *which meet at* $(\alpha,\beta) = (1,-1)$. *See Figure* 4.1. *At least one root satisfies* $\Re(z) > 0$ *for* (α,β) *belonging to the open complementary region on the right.*

Proof. Let's denote by AS the open connected region bounded by C_0 and part of the line $\beta = -\alpha$ in the (α,β) plane. The integer-valued function Z is continuous on AS by Lemma 4.3. Therefore, it must be constant by Lemma 4.3. Because $F(z,-1,0) = z+1 = 0$ if and only if $z = -1$, we know that $Z(-1,0) = 0$ so it follows that $Z = 0$ in AS. There are no purely imaginary roots or zero roots in AS, therefore we have shown that $\Re(z) < 0$ for every root when $(\alpha,\beta) \in AS$.

Consider the connected component of the complement of I bounded below by the line $\alpha + \beta = 0$ and bounded above by C_1. As $F(z,\alpha,0) = z - \alpha$ it follows that $Z(\alpha,0) = 1$ for $\alpha > 0$ and hence $Z = 1$ on this entire component.

Now we show that by crossing the boundary of AS through a point $(\alpha_0,\beta_0) = (\alpha(y_0),\beta(y_0))$ for some $y_0 \in (0,\pi)$ of C_0, away from $(1,-1)$, a conjugate pair of roots $\pm iy_0$ at (α_0,β_0) move into the right half-plane $\Re(z) > 0$. We use the implicit function theorem A.3 and the formula for α_0, β_0 in terms of y_0. We have $F(iy_0,\alpha_0,\beta_0) = 0$ and

$$F_z(iy_0,\alpha_0,\beta_0) = 1 + \beta_0 e^{-iy_0} = 1 - \frac{y_0}{\sin(y_0)}e^{-iy_0} = (1-\alpha_0) + iy_0 \neq 0$$

and

$$F_\alpha(iy_0,\alpha_0,\beta_0) = -1$$

so the implicit function theorem implies that the equation $F(z,\alpha,\beta_0) = 0$ (where β is fixed) has a solution $z = z(\alpha)$ for α near α_0 satisfying $z(\alpha_0) = iy_0$ and

$$\frac{dz}{d\alpha}(\alpha_0) = -\frac{F_\alpha}{F_z} = \frac{1}{(1-\alpha_0)+iy_0} = \frac{(1-\alpha_0)-iy_0}{(1-\alpha_0)^2+y_0^2}$$

So

$$z(\alpha) = iy_0 + (\alpha - \alpha_0)\frac{(1-\alpha_0)-iy_0}{(1-\alpha_0)^2+y_0^2} + O([\alpha-\alpha_0]^2)$$

and we see that

$$\Re(z(\alpha)) = (\alpha - \alpha_0)\frac{(1-\alpha_0)}{(1-\alpha_0)^2+y_0^2} + O([\alpha-\alpha_0]^2)$$

for α near α_0. Because $\alpha_0 < 1$, we see that $\Re(z(\alpha)) > 0$ for $\alpha > \alpha_0$. This proves that there is a complex conjugate pair of roots with positive real part just below but near C_0.

Z must be constant on the connected component D of the complement of I bounded above by C_0 and the portion of the line $\alpha + \beta = 0$ for $\alpha \geq 1$ and from below by C_2. It follows from the previous paragraph that $Z \geq 2$ on this region. We show that $Z = 2$ in this region. Suppose this is false and choose any point (α_0, β_0) on C_0 except for the corner point $(1, -1)$. Denote by iy_0 the unique purely imaginary root with $0 < y_0 < \pi$ such that $F(iy_0, \alpha_0, \beta_0) = 0$. If (α_n, β_n) is any sequence in D converging to (α_0, β_0) then there must exist a corresponding sequence of roots z_n, $F(z_n, \alpha_n, \beta_n) = 0$, with $\Re(z_n) > 0$ and z_n distinct from the complex conjugate pair constructed in the previous paragraph. By the a priori bounds obtained on roots with positive real part in Lemma 4.3, $\{z_n\}_n$ is a bounded sequence and so we may as well assume it converges to z_0, necessarily satisfying $\Re(z_0) \geq 0$ and $F(z_0, \alpha_0, \beta_0) = 0$. $z_0 \neq iy_0$ because in that case z_n would agree with the complex conjugate pair constructed in the previous paragraph. So $\Re(z_0) > 0$ and the implicit function theorem implies that there is a smooth family of roots $z = z(\alpha, \beta)$ for (α, β) near (α_0, β_0) satisfying $z(\alpha_0, \beta_0) = z_0$. But this contradicts that $Z = 0$ in AS.

Similar arguments to those in the previous paragraph show that $Z = 2n + 2$ between C_{2n} and C_{2n+2} and $Z = 2n + 3$ between C_{2n+1} and C_{2n+3} for $n = 0, 1, 2, \ldots$.
$\square$

Recall that when $\alpha = 0$ we showed in Proposition 2.1 that the stability region for β is given by $-\pi/2 < \beta < 0$ which corresponds to the segment on the β axis in the stability region Figure 4.1.

Now let's return to our original problem of determining the stability of the steady state $x = 0$ of scalar equation (4.18) which depends on characteristic equation (4.19). We assume that $A + B \neq 0$ for otherwise $\lambda = 0$ is a root.

Theorem 4.7 *The following hold for* (4.18):

(a) If $A + B > 0$, then $x = 0$ is unstable.
(b) If $A + B < 0$ and $B \geq A$, then $x = 0$ is asymptotically stable.
(c) If $A + B < 0$ and $B < A$, then there exists $r^* > 0$ such that $x = 0$ is asymptotically stable for $0 < r < r^*$ and unstable for $r > r^*$.

In case (c), there exist a pair of purely imaginary roots at

$$r = r^* = (B^2 - A^2)^{-1/2} \cos^{-1}(-A/B)$$

Proof. Consider (a). Then $(\alpha, \beta) = r(A, B)$ never meets AS or its boundary so there is a root $\lambda = r^{-1}z$ with positive real part for all $r > 0$.

Consider (b). In this case, $(\alpha, \beta) = r(A, B)$ lies entirely within AS for all $r > 0$ so all its roots $\lambda = r^{-1}z$ satisfy $\Re(\lambda) < 0$.

In case (c), the ray $(\alpha, \beta) = r(A, B)$, $r > 0$, belongs to AS for small $r > 0$ but meets C_0 at exactly one value of $r := r^*$ and leaves AS for $r > r^*$. The fact that it meets C_0 exactly once follows from a simple calculation

$$0 < \frac{d\beta}{d\alpha} = \frac{\frac{\sin(y)}{y} - \cos(y)}{1 - \cos(y)\frac{\sin(y)}{y}} < 1, \quad 0 < y < \pi$$

where we used the fact that $\sin(y)/y$ strictly decreases from one to zero on $(0, \pi)$

The stability assertions follow from Theorem 4.3. $\quad\Box$

4.6 Principle of Linearized Stability

Consider the nonlinear functional differential equation

$$x'(t) = f(x_t) \tag{4.23}$$

Then $x(t) = x_0 \in \mathbb{R}^n$, $t \in \mathbb{R}$ is a *steady-state solution* of (4.23) if and only if

$$f(\hat{x}_0) = 0$$

where $\hat{x}_0 \in C$ is the constant function equal to x_0:

If $x(t)$ is a solution of (4.23) and

$$x(t) = x_0 + y(t)$$

then $y(t)$ satisfies

$$y'(t) = f(\hat{x}_0 + y_t) \tag{4.24}$$

We want to understand the behavior of solutions of (4.23) that start near $\hat{x}_0$ and for this, it suffices to understand the behavior of solutions of (4.24) for solutions that start near $y = 0$. We assume that

$$f(\hat{x}_0 + \phi) = L(\phi) + g(\phi), \phi \in C \tag{4.25}$$

where $L : C \to \mathbb{R}^n$ is a bounded linear function and $g : C \to \mathbb{R}^n$ is "higher order" in the sense that

$$\lim_{\phi \to 0} \frac{|g(\phi)|}{\|\phi\|} = 0. \tag{4.26}$$

Of course, this means that for every $\mu > 0$, there exists $\delta > 0$ such that

$$\|\phi\| \leq \delta \implies |g(\phi)| \leq \mu \|\phi\|.$$

The linear system

$$z'(t) = L(z_t) \tag{4.27}$$

is sometimes called the linearized (or variational) equation about the equilibrium $\hat{x}_0$. We must view it on the complex space $C = C([-r, 0], \mathbb{C}^n)$.

The main result of this section is the following. See [41, 26] for the proof.

Theorem 4.8 *Let* $\Delta(\lambda) = 0$ *denote the characteristic equation corresponding to* (4.27) *and suppose that*

$$-\sigma := \max_{\Delta(\lambda)=0} \Re(\lambda) < 0.$$

Then $\hat{x}_0$ *is a locally asymptotically stable steady state of* (4.23). *In fact, there exists* $b > 0$ *such that*

$$\|\phi - \hat{x}_0\| < b \Longrightarrow \|x_t(\phi) - \hat{x}_0\| \le K\|\phi - \hat{x}_0\|e^{-\sigma t/2}, t \ge 0.$$

If $\Re(\lambda) > 0$ *for some characteristic root, then* $\hat{x}_0$ *is unstable.*

Consider the special case of (4.23)

$$x'(t) = F(x(t), x(t-r)) \tag{4.28}$$

where we assume that $F : D \times D \to \mathbb{R}^n$ is continuously differentiable and $D \subset \mathbb{R}^n$ is open. If $F(x_0, x_0) = 0$ for some $x_0 \in D$, then $x(t) = x_0$, $t \in \mathbb{R}$ is an equilibrium solution. Then $f(\phi) = F(\phi(0), \phi(-r))$ so (4.25) becomes

$$f(\hat{x}_0 + \phi) = A\phi(0) + B\phi(-r) + G(\phi(0), \phi(-r))$$

where $A = f_x(x_0, x_0)$, $B = f_y(x_0, x_0)$.

It follows that the linearized system about $x = x_0$ for (4.28) is

$$x'(t) = Ax(t) + Bx(t-r) \tag{4.29}$$

As an illustration of the use of Theorem 4.8, consider the scaled version of the delayed chemostat model (1.17)

$$S'(t) = 1 - S(t) - f(S(t))x(t) \tag{4.30}$$
$$x'(t) = e^{-r}f(S(t-r))x(t-r) - x(t)$$

where $f(S) = mS/(a+S)$ and $x, S \ge 0$. Nonnegative equilibria consist of the "washout state" $(S, x) = (1, 0)$ and, if $f(1) > e^r$, the "survival state" $(S, x) = (\bar{S}, \bar{x})$, where $f(\bar{S}) = e^r$ and $\bar{x} = (1 - \bar{S})e^{-r}$. The reader should verify that $\bar{x} > 0$. The matrices A and B, as in (4.29), at a generic point (S, x) are given by

$$A = \begin{pmatrix} -1 - xf'(S) & -f(S) \\ 0 & -1 \end{pmatrix}, B = e^{-r}\begin{pmatrix} 0 & 0 \\ f'(S)x & f(S) \end{pmatrix}$$

We find that (4.14) becomes

$$(\lambda + 1)(\lambda + 1 + xf'(S) - e^{-r}f(S)e^{-\lambda r}) = 0$$

What luck that it factors! Aside from the root $\lambda = -1$, the other factor leads to

$$\lambda = -1 - xf'(S) + e^{-r}f(S)e^{-\lambda r} \tag{4.31}$$

We use Theorem 4.7 but let's use a and b for the coefficients in (4.19) because A and B are in use here as matrices. For the washout state, we find that $a = -1$ and $b = e^{-r}f(1)$. Hence, $a + b = -1 + e^{-r}f(1)$ and $b > a$. Theorem 4.7 implies that the washout state is asymptotically stable when $e^{-r}f(1) < 1$ but unstable if $e^{-r}f(1) > 1$, when the survival state exists.

For the survival state, $a = -1 - \bar{x}f'(\bar{S})$ and $b = 1$, so $a + b < 0$ and $b > a$. Therefore, the survival state is asymptotically stable when it exists.

4.7 Absolute Stability

One often encounters the characteristic equation in the form

$$p(\lambda) + q(\lambda)e^{-r\lambda} = 0 \tag{4.32}$$

where p and q are polynomials with real coefficients and $r > 0$ is the delay. For example, the characteristic equation for a two-dimensional system, (4.14), has this property when $\det(B) = 0$, a case that is typical inasmuch as often an equation has only one delayed argument.

Typically, p has higher degree than q (see (4.14)). Brauer [8] proves the following result.

Proposition 4.9 *Let p, q be polynomials with real coefficients. Suppose*

(a) $p(\lambda) \neq 0$, $\Re(\lambda) \geq 0$.

(b) $|q(iy)| < |p(iy)|$, $0 \leq y < \infty$.

(c) $\lim_{|\lambda| \to \infty, \, \Re(\lambda) \geq 0} |q(\lambda)/p(\lambda)| = 0$.

Then $\Re(\lambda) < 0$ for every root λ and all $r \geq 0$.

The conclusion of Proposition 4.9 is called absolute stability, as stability holds for every value of the delay.

A simple corollary is stated and proved below.

Corollary 4.10 *Let p be a polynomial with real coefficients, and have leading coefficient one. Let $q = c$ be a constant. If*

(i) All roots of p are real and negative and $|p(0)| > |c|$, or
(ii) $p(\lambda) = \lambda^2 + a\lambda + b$, $a, b > 0$ and either

  - *$b > |c|$ and $a^2 \geq 2b$, or*
  - *$a\sqrt{(4b - a^2)} > 2|c|$ and $a^2 < 2b$,*

Then $\Re(\lambda) < 0$ for every root λ and all $r \geq 0$.

Proof. (i) If there is a root λ satisfying $\Re(\lambda) \geq 0$, then putting the exponential term on one side and the polynomial on the other and taking the modulus leads to

$$|p(\lambda)| = \prod_{i=1}^{n} |\lambda - \lambda_i| = |c|e^{-r\Re(\lambda)} \leq |c|, \qquad (4.33)$$

where λ_i are the roots of p. Obviously, $|\lambda - \lambda_i| \geq |\lambda_i|$ holds if $\lambda_i < 0$ and $\Re(\lambda) \geq 0$. Thus, $|p(0)| = \prod_i |\lambda_i| \leq |c|$ must hold. As this contradicts (i), the result follows.

To prove (ii), note that $|p(\lambda)| > 0$ for $\Re(\lambda) \geq 0$ because $a, b > 0$ imply its roots lie in the open left half-plane; also, $|p(\lambda)| \to \infty$ as $|\lambda| \to \infty$. Thus, $|p(\lambda)|$ must attain its minimum in $\Re(\lambda) \geq 0$ on the imaginary axis. We ask the reader to prove this in Exercise 4.15. The minimum value of

$$|p(iy)|^2 = |(b - y^2) + iay|^2 = (b - y^2)^2 + (ay)^2 = y^4 + (a^2 - 2b)y^2 + b^2,$$

is the minimum value of $g(x) = x^2 + (a^2 - 2b)x + b^2$ for $x \geq 0$. If $a^2 - 2b \geq 0$, then this minimum is $g(0) = b^2$, whereas if $a^2 - 2b < 0$, it is $g((2b - a^2)/2) = (4ba^2 - a^4)/4$. Our hypothesis that there is a root with $\Re(\lambda) \geq 0$ and (4.33) leads to $b \leq |c|$ in the case where $a^2 - 2b \geq 0$ and to $(a/2)\sqrt{(4b - a^2)} \leq |c|$ in the case where $a^2 - 2b < 0$. $\square$

It is interesting to compare the above result with Proposition 4.9 for

$$\lambda - \alpha = \beta e^{-r\lambda}$$

whose stability region is completely described in Proposition 4.6. Proposition 4.9 asserts that if $\alpha < 0$ and $|\alpha| < |\beta|$ then all roots satisfy $\Re(\lambda) < 0$ for every value of $r \geq 0$. This region of absolute stability is substantially smaller than the stability region for $r = 1$ shown there.

Exercises

Exercise 4.1. Let $f : [0, \infty) \to \mathbb{C}^n$ be continuous. Show that the initial-value problem

$$x'(t) = L(x_t) + f(t) \qquad (4.34)$$

and (4.4) have a unique solution defined for all $t \geq 0$

Exercise 4.2. Extend the superposition principle to (4.34) by showing that the solution of (4.34),(4.4) may be expressed as $x(t) = x(t; \phi, f) = x(t; \phi, 0) + x(t; 0, f)$ where $x(t; \phi, 0)$ is a solution of (4.1),(4.4) and $x(t; 0, f)$ is a solution of (4.34) corresponding to zero initial data.

Exercise 4.3. Verify that (4.5) defines a linear system in the sense of (4.1) by identifying the linear map L and showing that it is bounded. Hint: Use that $\int_0^{r_{ij}} |k_{ij}(s)| ds < \infty$.

Exercise 4.4. In taking the Laplace transform, we have tacitly assumed that the solution of (4.7) is exponentially bounded. Assume that f is exponentially bounded and continuous. Show that x is exponentially bounded as follows. Multiply (4.7) by e^{ct} where c is chosen such that $g(t) = f(t)e^{ct}$ satisfies $|g(t)| \leq Me^{-t}$ for some M and let $v(t) = x(t)e^{ct}$. Then v satisfies an equation of the same form as (4.7) but with inhomogeneous term f replaced by g. It suffices to show that v is exponentially bounded. After renaming variables, we return to Equation (4.7) where now $|f(t)| \leq Me^{-t}$. Define $u : [0, \infty) \rightarrow [0, \infty)$ by $u(t) = \max_{-r \leq s \leq t} |x(t)|$. Then $u(t)$ is continuous and nondecreasing. Starting from

$$x(t) = x_0 + \int_0^t Ax(\tau)d\tau + \int_0^t Bx(\tau - r)d\tau + \int_0^t f(\tau)d\tau$$

show that

$$|x(t)| \leq u(0) + L + (|A| + |B|) \int_0^t u(\tau)d\tau$$

where $L = \int_0^\infty |f(\tau)|d\tau$. Argue that this implies

$$u(t) \leq u(0) + L + (|A| + |B|) \int_0^t u(\tau)d\tau$$

and by Gronwall's inequality, we have

$$|x(t)| \leq u(t) \leq [u(0) + L]e^{(|A| + |B|)t}$$

Exercise 4.5. If A and B are 2×2 matrices, show that the characteristic equation can be written as

$$\lambda^2 - (tr(A))\lambda + \det(A) + e^{-2r\lambda} \det(B) + e^{-r\lambda}[C - \lambda(tr(B))] = 0$$

where

$$C = \det(a^1|b^2) + \det(b^1|a^2)$$

$\det(A)$ denotes the determinant of A, $tr(A)$ denotes its trace, and $(a^1|b^2)$ denotes the matrix with first column from A and second column from B.

Exercise 4.6. Determine an explicit form for the characteristic equation of (4.3) similar to (4.13).

Exercise 4.7. Show that $z = 0$ is a double root of (4.20) only when $\alpha = 1$ and $\beta = -1$. In fact, the only double roots are at $z = \alpha - 1$ when $\alpha = 1 + \log|\beta|$, $\beta < 0$. There are no roots of order three or higher.

Exercise 4.8. Show that the curves C_n, defined below (4.21), have the properties described. Show also that they are asymptotic to the lines $\beta = \pm\alpha$ as $y \rightarrow n\pi, (n+1)\pi$.

Exercise 4.9. We have implicitly assumed that coefficients A and B are independent of the delay r in Theorem 4.7. However, in applications it often happens that A and

B are functions of the delay r. Prove that part (a) of the theorem remains valid in this case by showing the existence of a positive characteristic root. Prove that part b remains valid in this case as well. Argue by contradiction, assuming there is a nonzero root λ with $\Re(\lambda) \geq 0$. Use that $|A| < |\lambda - A| = |B|e^{-r\Re(\lambda)} \leq |B|$ to obtain a contradiction.

Exercise 4.10. Find an expression for $g(\phi) = G(\phi(0), \phi(-r))$ and show that it satisfies (4.26).

Exercise 4.11. Find the equilibria for Nicholson's blowfly equation (1.3) and determine their stability properties.

Exercise 4.12. Consider the Lotka-Volterra competition system

$$x'(t) = x(t)[2 - ax(t) - by(t - r)]$$
$$y'(t) = y(t)[2 - cx(t - r) - dy(t)]$$

(a.) Determine the stability of the positive steady state when $a = d = 2$ and $b = c = 1$. How does it depend on r?
(b.) Determine the stability of the positive steady state when $a = d = 1$ and $b = c = 2$. How does it depend on r?

Exercise 4.13. Find the steady state with positive components for the delayed predator-prey model

$$x'(t) = x(t)(m - x(t) - y(t))$$
$$y'(t) = y(t)(-1 + ax(t - r))$$

where $m, r, a > 0$. Find the linear variational equation and the associated characteristic equation.

Exercise 4.14. J.D. Murray in Chapter 6, Section 6, of his famous book [61] introduced the system (1.16). He claims that the positive steady state can be made unstable if the delay r is large enough. Is he correct? Determine the steady state and the characteristic equation. Hint: Try not to use the explicit formula for $f(T)$ but only the qualitative shape of its graph to determine when there is a steady state. You do not need the exact formula for the steady-state value of T.

Exercise 4.15. In the proof of Corollary 4.10(ii), we claimed that $|p(\lambda)|$ attains its minimum over the region $\Re(\lambda) \geq 0$ on the imaginary axis. Prove this assertion. Hint: One way is to set the first partial derivatives $(|p(\lambda)|^2)_x = (|p(\lambda)|^2)_y = 0$, where $\lambda = x + iy$. In doing so, use only that p is an analytic function so the Cauchy–Riemann equations (A.1) hold. Show that if there were a solution, then as $|p(\lambda)| > 0$ it must follow that $p'(\lambda) = 0$. But $p'(\lambda) = 2\lambda + a$ has no zero in $\Re(\lambda) > 0$.

Chapter 5
Semidynamical Systems and Delay Equations

Abstract An autonomous system of delay differential equations is shown to generate a semidynamical system on the space C of continuous functions on the delay interval. Omega limit sets are defined and shown to have the same properties as for ODEs, with minor exceptions, although they are subsets of C. The dynamics of the delayed logistic equation and the chemostat model are treated in detail. A special class of delay equations is shown to generate monotone dynamics; solutions converge to equilibrium. Liapunov functions and the LaSalle invariance principle are used to study the dynamics of a delayed logistic equation with both instantaneous and delayed density dependence.

5.1 The Dynamical Systems Viewpoint

A dynamical system consists of a set X of "states" and a rule Λ describing how states change with time. X is called the state space. If at time s you are at state x and later, at time t, find yourself at state x', then $x' = \Lambda(t,s,x)$. The transition rule (function) should take as inputs the "initial" time s and state x and produce the new state x' at time t. The function Λ should have some properties consistent with this interpretation of its meaning. For example, it should obviously satisfy

$$\Lambda(s,s,x) = x, \forall s,x \tag{5.1}$$

If at time s we are at state x, at time r we are at state x', and at time t are at state x'', then $x' = \Lambda(r,s,x)$, $x'' = \Lambda(t,r,x')$, and $x'' = \Lambda(t,s,x)$ should hold in keeping with our interpretation. It follows that Λ should satisfy

$$\Lambda(t,s,x) = \Lambda(t,r,\Lambda(r,s,x)), \forall t,s,r,x \tag{5.2}$$

So far, our arguments have been informal and, in particular, we have not specified from what set one should take the "time" t. For discrete-time dynamical systems, we might choose this set to be the integers $\mathbb{Z}$, or the nonnegative integers $\mathbb{Z}_+$. If, for

H. Smith, *An Introduction to Delay Differential Equations with Applications to the Life Sciences,*
Texts in Applied Mathematics 57, DOI 10.1007/978-1-4419-7646-8_5,
© Springer Science+Business Media, LLC 2011

example, we are given a sequence of maps $F_n : X \to X$ for $n \in \mathbb{Z}$ we may consider
the dynamics generated by the recursion

$$x_{n+1} = F_n(x_n), n \geq s, x_s = x$$

where $s \in \mathbb{Z}$ is the initial time and x the initial state. This generates the sequence:

$$x_s = x \to x_{s+1} = F_s(x_s) \to x_{s+2} = F_{s+1}(x_{s+1}) \to \cdots$$

The map F_s is applied at time s to get the state at time $s + 1$. The transition rule is

$$\Lambda(t,s,x) = F_{t-1} \circ F_{t-2} \circ \cdots \circ F_s(x)$$

where $\circ$ denotes function composition.

If we are given a single map $F : X \to X$ we may consider the dynamics generated
by

$$x_{n+1} = F(x_n), n \geq s, \ x_s = x$$

which is formally the same as above where F_n is the constant sequence $F_n = F$.
Define $F^{(p)}(x) = (F \circ F \circ F \circ \cdots \circ F)(x)$ to be the p-fold composition of F with
itself where, in general, $p \in \mathbb{Z}_+$. Then we define $\Lambda(t,s,x) = F^{(t-s)}(x)$ for $x \in X$ and
$s \in \mathbb{Z}$ but $t \in \mathbb{Z}$ must in general satisfy $t > s$ because F need not be invertible; if F
is invertible, then no restriction on t is necessary. A noninvertible map F generates
the simplest dynamical system where one sees clearly why we can generally not go
backward in time.

For continuous-time dynamical systems, one may choose the reals or the nonnegative reals. The quintessential example of a dynamical system is that generated by
a system of ODEs. Here, we typically are interested in solutions of the initial-value
problem

$$x' = f(t,x), x(s) = x_0$$

Under suitable conditions, there is a unique solution $x(t)$, which we often write as
$x(t,s,x_0)$ to remind ourselves that the solution depends on all three arguments. In
this case, $\Lambda(t,s,x_0) = x(t,s,x_0)$. For ODEs there is no asymmetry between the past
and the future so it is natural to take the real line as our time set.

We have noted already that delay differential equations can generally be solved
only forward in time, and that we should expect a continuous-time dynamical system
in this case. Therefore, we specialize our formal definition of a dynamical system
with these features in mind. Define

$$S = \{(t,s) \in \mathbb{R} \times \mathbb{R} : t \geq s\}$$

We say that $\Lambda : S \times X \to X$ is a *semidynamical system* if it satisfies (5.1) and (5.2);
for the latter, (t,r) and (r,s) must belong to S. The "semi" in semidynamical reflects
the restriction that we may only go forward in time, that is, $t \geq s$ in the definition
of $\Lambda(t,s,x)$. In defining a dynamical system, we replace S above by $S = \mathbb{R} \times \mathbb{R}$ and

require (5.1) and (5.2) to hold without restriction. From a dynamical system, we obtain a semidynamical system by restriction of the domain.

In practice, we demand that our semidynamical system has some continuity properties but it was useful to start out by ignoring this to call attention to the "algebraic" conditions (5.1) and (5.2). Thus, hereafter, we assume that our state space X is a metric space (X, d) with metric d and that Λ is continuous: if $(t, s) \in S$, $x \in X$ and if $\{(t_n, s_n)\}_n$ is a sequence in S and $\{x_n\}_n$ is a sequence in X such that $(t_n, s_n) \to (t, s)$ and $x_n \to x$, then $\Lambda(t_n, s_n, x_n) \to \Lambda(t, s, x)$.

It is useful to have the notion of a *solution* for a semidynamical system. It should be a continuous function $\sigma : I \to X$, where $I \subset \mathbb{R}$ is an interval containing more than a single point, tracing out the states followed by the system. By this we mean that if $s, t \in I$ and $t \geq s$ then $\sigma(t) = \Lambda(t, s, \sigma(s))$. The reader should verify that $\sigma : [s, \infty) \to X$ defined by $\sigma(t) = \Lambda(t, s, x)$ is a solution.

Let $\sigma : I \to X$ and $v : J \to X$ be two solutions. If $s \in I \cap J$ and $\sigma(s) = v(s)$, then $\sigma(t) = v(t)$ for all $t > s$ belonging to $I \cap J$. See the exercises.

A semidynamical system is said to be *autonomous* if Λ satisfies:

$$\Lambda(t, s, x) = \Lambda(t + r, s + r, x), \forall (t, s) \in S, r \in \mathbb{R}, x \in X \tag{5.3}$$

In that case, by taking $r = -s$, we find that $\Lambda(t, s, x) = \Lambda(t - s, 0, x)$; the initial time is irrelevant and only the elapsed time $t - s$ matters. In the case where Λ is an autonomous semidynamical system, we define $\Phi : \mathbb{R}_+ \times X \to X$ by $\Phi(t, x) = \Lambda(t, 0, x)$ so we have $\Lambda(t, s, x) = \Phi(t - s, x)$. The map Φ completely specifies Λ. It is easy to verify that (5.1) and (5.2) imply that Φ satisfies:

$$\Phi(0, x) = x, \Phi(t, \Phi(s, x)) = \Phi(t + s, x), t, s \geq 0, x \in X \tag{5.4}$$

Lemma 5.1. *Semidynamical system Λ is autonomous if and only if whenever $\sigma(t)$ is a solution defined on interval I and $c \in \mathbb{R}$, then $\sigma(t + c)$ is a solution on interval $I - c$, where $I - c = \{t - c : t \in I\}$.*

Proof. Suppose that Λ is autonomous, $\sigma : I \to X$ is a solution, $c \in \mathbb{R}$, and $u(t) = \sigma(t + c)$, defined for $t \in I - c$. If $t, s \in I - c$ satisfy $t \geq s$ then $t = t' - c, s = s' - c$ where $s', t' \in I$ and $t' \geq s'$. Consequently,

$$\Lambda(t, s, u(s)) = \Lambda(t' - c, s' - c, \sigma(s')) = \Lambda(t', s', \sigma(s')) = \sigma(t') = u(t)$$

Therefore, $u(t)$ is a solution as required.

Conversely, suppose translates of solutions are solutions on the translated interval. We show that (5.3) holds. Let $\sigma : [s + r, \infty) \to X$ be defined by $\sigma(t) = \Lambda(t, s + r, x)$. We claim that it is a solution. If $t' \geq s' \geq s + r$ then

$$\Lambda(t', s', \sigma(s')) = \Lambda(t', s', \Lambda(s', s + r, x)) = \Lambda(t', s + r, x) = \sigma(t')$$

It follows that σ is a solution and hence so is $u(t) = \sigma(t + r)$, $t \geq s$ by hypothesis. But $u(s) = x$ which agrees with the solution $v(t) = \Lambda(t, s, x)$, $t \geq s$ at $t = s$. It follows

that $u(t) = v(t)$, $t \geq s$, or $\Lambda(t+r,s+r,x) = \Lambda(t,s,x)$, $t \geq s$, proving that (5.3) holds.
$\square$

See [66] for a thorough treatment of semidynamical systems.

5.2 Semiflows and Omega Limit Sets

In these notes we primarily focus on autonomous semidynamical systems. We call
the continuous map $\Phi : \mathbb{R}_+ \times X \to X$ a semiflow if it satisfies (5.4).

The *forward orbit* of a state x is defined by

$$O_+(x) = \{\Phi(t,x) : t \geq 0\}$$

An *equilibrium* is a state $e \in X$ that does not change with time: $\Phi(t,e) = e$ for all
$t \geq 0$, or equivalently, $O_+(e) = \{e\}$. If e is an equilibrium, then $\sigma : \mathbb{R} \to X$ defined
by $\sigma(t) = e$ is a solution.

The *omega limit set* is defined in the usual way:

$$\omega(x) = \{y \in X : \exists\{t_n\}_{n \geq 1}, \ t_n \to \infty, \ \Phi(t_n,x) \to y\}$$

A subset $A \subset X$ is *positively invariant* if $a \in A \Rightarrow O_+(a) \subset A$. It is *invariant* if
$\Phi(t,A) = A$ for all $t \geq 0$. Notice that this means A is positively invariant and that for
any $a \in A$ and $t \geq 0$, there exists $a' \in A$ such that $\Phi(t,a') = a$.

Now we can state the main result of this section.

Theorem 5.1 *The omega limit set $\omega(x)$ is closed and positively invariant. If $\overline{O_+(x)}$
is compact in X, then $\omega(x) \neq \emptyset$ and it is compact, connected, invariant, and*

$$\Phi(t,x) \to \omega(x), t \to \infty.$$

*This means that for $\varepsilon > 0$ there exists $T > 0$ such that if $t > T$ there exists $y \in \omega(x)$
such that $d(\Phi(t,x),y) < \varepsilon$.*

Proof. If $y \in \omega(x)$ there exists $t_n \uparrow \infty$ such that $\Phi(t_n,x) \to y$. If $t > 0$ then by continuity of Φ and (5.4), $\Phi(t,\Phi(t_n,x)) \to \Phi(t,y)$ and $\Phi(t+t_n,x) \to \Phi(t,y)$. Since
$t+t_n \uparrow \infty$ we see that $\Phi(t,y) \in \omega(x)$. This proves positive invariance of $\omega(x)$.

To see that $\omega(x)$ is closed, suppose that $\{y_n\}$ is a sequence of points of $\omega(x)$
converging to y. We must show that $y \in \omega(x)$. As $y_1 \in \omega(x)$ we may find $t_1 > 1$
such that $d(\Phi(t_1,x),y_1) < 1$. For similar reasons, we find $t_2 > \max\{t_1,2\}$ such that
$d(\Phi(t_2,x),y_2) < 1/2$. Proceeding inductively, for each $n \in \mathbb{N}$ we find $t_n > t_{n-1},n$
such that $d(\Phi(t_n,x),y_n) < 1/n$. Then $d(\Phi(t_n,x),y) \leq d(\Phi(t_n,x),y_n) + d(y_n,y) \to 0$
implying that $y \in \omega(x)$.

If $\overline{O_+(x)}$ is compact in X and $t_n \uparrow \infty$ then $\{\Phi(t_n,x)\} \subset \overline{O_+(x)}$ so it must have a
convergent subsequence. Obviously, the limit of this sequence belongs to $\omega(x)$ so
the latter is nonempty. Inasmuch as $\omega(x)$ is a closed subset of the compact set $\overline{O_+(x)}$

it is compact. To establish invariance, given $y \in \omega(x)$ and $s > 0$ we want to show that there exists $z \in \omega(x)$ such that $\Phi(s,z) = y$. There exists $t_n \uparrow \infty$ such that $\Phi(t_n,x) \to y$. $\{\Phi(t_n - s,x)\} \subset \overline{O_+(x)}$ has a convergent subsequence so we may assume, by renaming our convergent subsequence, that $\Phi(t_n - s,x) \to z$ for some $z \in X$. Now using (5.4) and continuity as above we see that $\Phi(s,z) = \lim_{n\to\infty} \Phi(s, \Phi(t_n - s,x)) = \lim_{n\to\infty} \Phi(t_n,x) = y$.

To show that $\Phi(t,x) \to \omega(x)$, $t \to \infty$, we argue by contradiction. If it were false then for some $\varepsilon > 0$ there is, for each natural number $T = n$, some $t_n > n$ such that $d(\Phi(t_n,x),y) \geq \varepsilon$ for every $y \in \omega(x)$. But $\{\Phi(t_n,x)\}$ has a convergent subsequence so, on renumbering this subsequence, we may assume that $\Phi(t_n,x) \to w$ and obviously, $w \in \omega(x)$. Taking the limit in the above inequality, we find that $d(w,y) \geq \varepsilon$ for every $y \in \omega(x)$, including $y = w$ itself.

If $\omega(x)$ were disconnected then $\omega(x) = A \cup B$ where A and B are nonempty closed sets with $A \cap B = \emptyset$. It follows that A and B are compact and that we may find $\varepsilon > 0$ such that $U = \{x \in X : \exists y \in A \text{ such that } d(x,y) < \varepsilon\}$ and $V = \{x \in X : \exists y \in B \text{ such that } d(x,y) < \varepsilon\}$ are disjoint. By the previous paragraph, $\exists T > 0$ such that $t > T$ implies there is some $w \in \omega(x)$ such that $d(w, \Phi(t,x)) < \varepsilon$. Although w is not uniquely determined by t, the set A or B to which it belongs is uniquely defined. Let $I = \{t > T : w \in A\}$ and $J = \{t > T : w \in B\}$ where w corresponds to t as above. Then $(T,\infty) = I \cup J$ and $I \cap J = \emptyset$. It is not hard to see that I and J are nonempty open sets. This implies that (T,∞) is disconnected, a contradiction. $\square$

Remark 5.2 *According to Exercise 5.6 and Theorem 5.1, if $\overline{O_+(x)}$ is compact then through each $y \in \omega(x)$ there exists at least one solution $\sigma : \mathbb{R} \to \omega(x)$ satisfying $\sigma(0) = y$.*

5.3 SemiDynamical Systems Induced by Delay Equations

Consider the initial value problem for the time-dependent delay differential equation

$$x'(t) = f(t,x_t), x_s = \phi \tag{5.5}$$

where $s \in \mathbb{R}$, f is continuous, and $\phi \in C$. Here, we assume that there is a unique solution defined for all $t \geq s$ for every such initial-value problem $(s,\phi) \in \mathbb{R} \times C$. Of course, this is hard to verify in applications, but such is required if we are to show that (5.5) is to generate a semidynamical system. Write $x(t,s,\phi)$ for this solution defined for $t \geq s$. The state of the system at time t is $x_t(s,\phi) \in C$, defined in the usual way as $x_t(s,\phi)(\theta) = x(t+\theta,s,\phi)$, $-r \leq \theta \leq 0$. In an analogous fashion as for systems of ODEs, we show that

$$\Lambda(t,s,\phi) = x_t(s,\phi) \tag{5.6}$$

defines a semidynamical system on C.

Proposition 5.3 Λ *defines a semidynamical system on C.*

Proof. We must show that $x_t(s, \phi) = x_t(r, x_r(s, \phi))$ for $t \geq r \geq s$, $\phi \in C$. Equality clearly holds when $t = r$. Both $x(t, s, \phi)$ and $x(t, r, x_r(s, \phi))$ are solutions of (5.5) on $t \geq r$ and their respective states in C at $t = r$ are the same, so by uniqueness of solutions to initial-value problems for (5.5), they agree for $t \geq r$.

The continuity of $\Lambda(t, s, \phi) = x_t(s, \phi)$ in all arguments follows directly from Theorem 2.2, Chapter 2 in [41]. In the case where f satisfies the Lipschitz condition on bounded sets (3.13), then it also follows from Theorem 3.7.   □

Having shown that (5.5) defines a semidynamical system on C, we now briefly turn to reconciling two definitions of "solution". Recall, we have defined what we mean by a solution of (5.5) in Chapter 2. In a previous section, we defined what is meant by a solution of the semidynamical system Λ. Proposition 5.3 shows that solutions of the initial-value problems (5.5) define the semidynamical system and this implies that solutions of the delay equation correspond to solutions of the semidynamical system Λ. Below we show the converse.

Lemma 5.2. *Let* $\sigma : I \to C$ *be a solution of the semidynamical system* (5.6) *and let* $y : I \to \mathbb{R}^n$ *be defined by* $y(t) = \sigma(t)(\theta = 0)$. *Then* $y(t)$ *is a solution of* $x'(t) = f(t, x_t)$ *for* $t \in I$. *In the case where* $I = [a, b]$ *or* $I = [a, b)$ *for some* $b \leq \infty$, *then* $\sigma(t) = x_t(a, \sigma(a))$ *so* y *can be continuously extended to the interval* $[a - r, a)$ *by* $y_a = \sigma(a)$.

Proof. Let s be any interior point of I. Then we have

$$\sigma(t) = \Lambda(t, s, \sigma(s)) = x_t(s, \sigma(s)), t \geq s, t \in I.$$

Evaluating at $\theta = 0$, we find that $y(t) = x(t, s, \sigma(s))$, the unique solution of $x'(t) = f(t, x_t)$ satisfying $x_s = \sigma(s)$. It follows that $y(t)$ is a solution for $t \geq s$. As s was chosen arbitrarily, we conclude that $y(t)$ is a solution of $x'(t) = f(t, x_t)$ for all $t \in I$. In the case $I = [a, b]$ or $I = [a, b)$, we may take $s = a$ and the final assertion follows immediately.   □

We now turn to the study of the time-independent special case of Equation (5.5):

$$x'(t) = f(x_t) \tag{5.7}$$

Exercise 5.7, Lemma 5.2, and Lemma 5.1 imply that the semidynamical system Λ generated by (5.7) according to (5.6) is autonomous:

$$\Lambda(t, s, \phi) = x_t(s, \phi) = x_{t-s}(0, \phi)$$

Because $x(t, s, \phi) = x(t - s, 0, \phi)$ for solutions of (5.7), we may as well consider the initial-value problem with $s = 0$ because we may always translate time to make this so. At the same time, we simplify our notation by writing $x(t, \phi) = x(t, 0, \phi)$ and $x_t(\phi) = x_t(0, \phi)$. Our autonomous semidynamical system, or semiflow for short, can now be expressed as

$$\Phi(t, \phi) = x_t(\phi), t \geq 0, \phi \in C \tag{5.8}$$

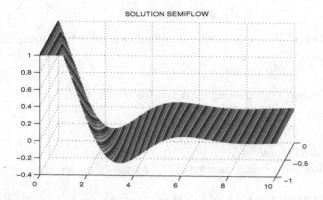

Fig. 5.1 Solution semiflow for $x'(t) = -0.75x(t-1)$ with $\phi = \hat{1}$.

The semiflow generated by $x'(t) = -0.75x(t-1)$ with $x_0 = \hat{1}$ may be visualized as in Figure 5.1 plotting the surface $(t, \theta) \to x(t+\theta)$ for $-1 \le \theta \le 0$, $0 \le t \le 10$. The surface is foliated by curves, each of which is a computer approximation to x_t for some $t \in [0, 10]$.

Keep in mind that we are assuming that for each $\phi \in C$, there exists a unique solution $x(t, \phi)$ of the initial-value problem (5.7) and $x_0 = \phi$ and that it extends to the entire half-line $[0, \infty)$. The orbit of ϕ is

$$O_+(\phi) = \{x_t(\phi) : t \ge 0\} \subset C$$

Recall $e \in C$ is an equilibrium if its orbit is a singleton: $O_+(e) = \{e\}$. This just means that e is constant as expected.

Proposition 5.4 *e is an equilibrium if and only if e is a constant function satisfying* $f(e) = 0$. *In that case,* $x(t) = e(0)$, $t \in \mathbb{R}$ *is a solution of* (5.7).

Proof. If e is an equilibrium then $x_t(e) = e$ for all $t \ge 0$ so $x(t+\theta, e) = e(\theta)$ for all $t \ge 0$ and $\theta \in [-r, 0]$. Putting $\theta = 0$ into this gives $x(t, e) = e(0)$ for all $t \ge 0$ so $x(t)$ is constant, $x'(t) = 0$, and therefore $f(e) = 0$. Putting $t = -\theta$ into the same equality gives $e(0) = e(\theta)$ for all θ so e is a constant function. The converse is trivial (by uniqueness of solutions). $\square$

There is a natural way to assign to each delay differential equation an ordinary differential equation that has the same equilibria. It is called "ignoring the delays." In order to describe the correspondence, it is useful to have notation for the natural embedding $\mathbb{R}^n \to C = C([-r, 0], \mathbb{R}^n)$ given by $x \to \hat{x}$ where $\hat{x}(\theta) = x$, $-r \le \theta \le 0$. Given $f : C \to \mathbb{R}^n$ in the right-hand side of (5.7), define $F : \mathbb{R}^n \to \mathbb{R}^n$ by

$$F(x) = f(\hat{x}), x \in \mathbb{R}^n$$

Then the ODE

$$x' = F(x)$$

has the same equilibria as (5.7). For example, corresponding to the delay differential equation

$$N'(t) = N(t)[b - aN(t - r)]$$

where $f(\phi) = \phi(0)[b - a\phi(-r)]$, ignoring the delays leads to the ODE

$$N'(t) = N(t)[b - aN(t)]$$

where $F(N) = f(\hat{N}) = N[b - aN]$

Our earlier definition of the omega limit set of $O_+(\phi)$ may be expressed as

$$\omega(\phi) = \{\psi \in C : x_{t_n}(\phi) \to \psi \text{ some } t_n \uparrow \infty\} \tag{5.9}$$

Recall that we say the sequence $\{\phi_n\}$ in C converges to ϕ, and write: $\phi_n \to \phi$, when $\|\phi_n - \phi\| \to 0$, that is, when $\phi_n(\theta) \to \phi(\theta)$ uniformly on $[-r, 0]$.

According to Theorem 5.1, to show that an omega limit set $\omega(\phi)$ is nonempty we must show that the closure of the orbit $O_+(\phi)$ is compact. For ODEs this is easy, we just show that the orbit is bounded. The Heine–Borel theorem [65] says every closed and bounded subset of $\mathbb{R}^n$ is compact. But things are not so easy for delay differential equations. The Heine–Borel theorem does not hold for C! Of course, that is the whole point of the Ascoli–Arzela theorem A.1; we would have no need for it if closed and bounded subsets of C are necessarily compact.

Another consequence of the failure of the Heine–Borel theorem to hold for C is the following: if $A \subset C$ is closed and bounded, we cannot conclude from this that $f(A)$ is bounded in $\mathbb{R}^n$, where $f : C \to \mathbb{R}^n$ is continuous and defines our delay equation (5.7). If, however, f satisfies a Lipschitz condition on each bounded subset of C (3.13), then we can conclude that $f(A)$ is bounded in $\mathbb{R}^n$ whenever A is bounded in C. More generally, we say $f : C \to \mathbb{R}^n$ is *completely continuous* if it maps bounded sets in C to bounded sets in $\mathbb{R}^n$.

These two issues concerning Heine–Borel are related as we see in the following result which gives sufficient conditions for the compactness condition required in Theorem 5.1.

Proposition 5.5 *If f is completely continuous and $O_+(\phi)$ is bounded, then $\overline{O_+(\phi)}$ is compact in C.*

Proof. There is an $M > 0$ such that $\|x_t(\phi)\| \leq M$ for $t \geq 0$ and because f is completely continuous there is an $L > 0$ such that $|x'(t, \phi)| = |f(x_t)| \leq L$ for $t \geq 0$. So for $t \geq r$, $x_t(\phi)$ is continuously differentiable with derivative bounded in norm by L implying its Lipschitz constant is L. Thus $\{x_t(\phi) : t \geq r\}$ is a uniformly bounded and equicontinuous family of functions in C. Let $\{x_{t_n}(\phi)\}$ be a sequence in $O_+(\phi)$. $\overline{O_+(\phi)}$ is compact if we show that there is a convergent subsequence. There are two cases. If the sequence $\{t_n\}$ has a convergent subsequence $\{t_{n_k}\}$ with $t_{n_k} \to t_0$ then $x_{t_{n_k}} \to x_{t_0}$ as $k \to \infty$ by Lemma 3.1. If $\{t_n\}$ has no convergent subsequence, then $t_n \to \infty$. Because $\{x_{t_n}\}$ is a uniformly bounded and equicontinuous sequence, it has a uniformly convergent subsequence by the Ascoli–Arzela theorem A.1. □

As an immediate consequence of the previous result and Theorem 5.1, we have the following.

Corollary 5.6 *Let f in (5.7) be completely continuous and suppose $O_+(\phi)$ is bounded in C. Then $\omega(\phi)$ is nonempty, compact, connected, invariant, and $x_t(\phi) \to \omega(\phi)$ as $t \to \infty$.*

The reader should observe that boundedness of $O_+(\phi)$ in C is equivalent to the boundedness of $\{x(t, \phi) : t \geq 0\}$ in $\mathbb{R}^n$. The latter is just what we require for ODEs.

Some elementary applications of invariance of the omega limit set follow.

Proposition 5.7 *Suppose $\lim_{t \to \infty} x(t, \phi) = c$ for some constant c. If $\hat{c}$ denotes the element of C identically equal to the value c, then $\hat{c}$ is an equilibrium, $f(\hat{c}) = 0$, and, of course, $\omega(\phi) = \{\hat{c}\}$.*

Proof. It's easy to see that $x_t(\phi) \to \hat{c}$ as $t \to \infty$ so $\omega(\phi) = \{\hat{c}\}$. Because omega limit sets are invariant sets $O_+(\hat{c}) = \{\hat{c}\}$ for all $t \geq 0$ so it's an equilibrium. $\square$

Remark 5.8 *We have implicitly assumed that our semiflow is defined on all of C merely for ease of exposition. It is rarely the case in applications. Often, we may identify some positively invariant subset D of C, such as the nonnegative functions, and restrict our attention to initial data in this subset. In that case, we may define our semiflow Φ only on D. All of the above results carry over in a natural fashion if D is closed in C; if D is not closed, it may happen that omega limit points don't belong to D.*

As an application, consider the equation

$$x'(t) = -x(t-1)[1 - x^2(t)], t \geq 0. \tag{5.10}$$

It has three equilibria, the constant functions $-1, 0, 1$.

Let $D \subset C$ consisting of $\phi \in C$ satisfy $-1 \leq \phi(0) \leq 1$ and let $x(t, \phi)$ be the solution to the initial-value problem $x_0 = \phi$. Because

$$[1 - x(t)]' = x(t-1)(1 + x(t))[1 - x(t)]$$

we may conclude that

$$[1 - x(t)] = [1 - \phi(0)] \exp\left(\int_0^t x(s-1)(1 + x(s))ds\right) \geq 0$$

Inasmuch as $[1 + x(t)]$ satisfies a similar linear equation, we conclude that

$$[1 + x(t)] = [1 + \phi(0)] \exp\left(-\int_0^t x(s-1)(1 - x(s))ds\right) \geq 0$$

Therefore, we see that

$$-1 \leq x(t, \phi) \leq 1, t \geq 0$$

and hence the solution can be extended to all of $[0, \infty)$ by a standard argument and it's bounded. As $f(\phi) = \phi(-1)[1 - \phi^2(0)]$ is easily seen to be completely continuous, Corollary 5.6 implies that $\omega(\phi)$ is nonempty and has all the usual properties.

5.4 Monotone Dynamics

Consider the delay differential equation

$$x'(t) = f(x(t), x(t-r)) \tag{5.11}$$

where $f : \mathbb{R}^2 \to \mathbb{R}$ and f_x, f_y are continuous and

$$f_y(x, y) \geq 0 \tag{5.12}$$

Recall that (5.12) implies that the comparison Theorem 3.6 applies. As a consequence, the semiflow induced by (5.11) is monotone in the following sense.

Theorem 5.9 *Let $x(t)$ and $y(t)$ be two solutions of (5.11) defined on $[-r, T]$ for some $T > 0$. If $x_0 \leq y_0$, then $x_t \leq y_t$ for $0 \leq t \leq T$.*

Proof. This is an immediate consequence of Theorem 3.6. □

Remarkably, the dynamics of (5.11) shares many of the simple features as that of the ODE

$$x'(t) = f(x(t), x(t)) \tag{5.13}$$

Obviously, they share the same equilibria. We show that the stability properties of equilibria are generically the same. First, we establish a useful convergence result. Recall that if $a \in \mathbb{R}$, then $\hat{a} \in C$ is the constant function identically equal to a.

Proposition 5.10 *Suppose that there exists $a \in \mathbb{R}$ such that $f(a, a) \geq 0$. Then*

$$\phi \geq \hat{a} \Rightarrow x(t, \phi) \geq x(t, \hat{a}) \geq a \tag{5.14}$$

Moreover, the solution $t \to x(t, \hat{a})$ is nondecreasing on its maximal interval of existence; if it is bounded above, then it converges to an equilibrium b satisfying $a \leq b$. An analogous result holds, with inequalities reversed.

Proof. Change variables by $y = x - a$ yielding the equation $y'(t) = g(y(t), y(t-r))$ where $g(x, y) = f(a+x, a+y)$. g satisfies the hypotheses of Theorem 3.4: if $y \geq 0$ and $x = 0$, then $g(0, y) = f(a, a+y) \geq f(a, a) \geq 0$. Therefore, positivity is preserved by the transformed system in the sense that if $\phi \geq 0$, then $y(t, \phi) \geq 0$, $t \geq 0$. Clearly this implies that if $\phi \geq \hat{a}$, then $x(t, \phi) \geq a$, $t \geq 0$. In particular, $x(t, \hat{a}) \geq a$, $t \geq 0$, which in turn implies that $x_t(\hat{a}) \geq \hat{a}$, $t \geq 0$. By monotonicity of the semiflow via Theorem 5.9, we have that $x_{s+t}(\hat{a}) = x_s(x_t(\hat{a})) \geq x_s(\hat{a}) \geq \hat{a}$ for $s, t \geq 0$. This implies that

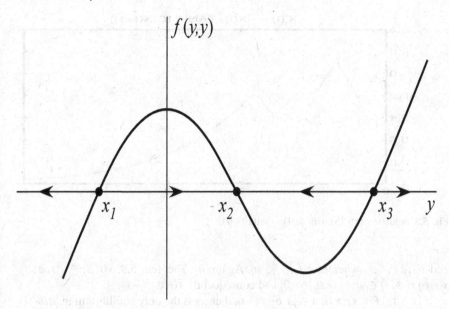

Fig. 5.2 Dynamics of (5.13).

$$0 \le t_1 \le t_2 \Rightarrow \hat{a} \le x_{t_1}(\hat{a}) \le x_{t_2}(\hat{a})$$

which immediately implies the asserted monotonicity of $t \to x(t, \hat{a})$. Convergence of $x(t, \hat{a})$ to equilibrium, if it is bounded above, follows from Proposition 5.7. Equation (5.14) follows directly from Theorem 5.9 because $\phi \ge \hat{a}$. $\square$

An immediate corollary of Proposition 5.10 and Theorem 5.9 is that the stability properties of (5.13) and of (5.11) are the same.

Corollary 5.11 *Let $x_0 \in (a,b)$ be an equilibrium of (5.13), and hence of (5.11). Suppose that*

$$(x - x_0) f(x,x) < 0, x \in [a,b], x \ne x_0$$

If $\phi(s) \in [a,b], s \in [-r,0]$, then $x(t,\phi) \to x_0$, $t \to \infty$.
Alternatively, suppose that

$$(x - x_0) f(x,x) > 0, x \in [a,b], x \ne x_0$$

If $\phi(s) \in [a,x_0)$, $s \in [-r,0]$, then $x(t,\phi) < a$ for some $t > 0$; If $\phi(s) \in (x_0,b]$, $s \in [-r,0]$, then $x(t,\phi) > b$ for some $t > 0$.

Proof. If $(x - x_0) f(x,x) < 0$, $x \in [a,b], x \ne x_0$ then x_0 is the only equilibrium in $[a,b]$ and $f(a,a) > 0 < f(b,b)$. As $\hat{a} < \hat{x}_0 < \hat{b}$, we conclude from Theorem 5.9 that $x(t,\hat{a}) \le x_0 \le x(t,\hat{b})$, $t \ge 0$. By Proposition 5.10,

$$a \le x(t,\hat{a}) \le x_0 \le x(t,\hat{b}) \le b, t \ge 0$$

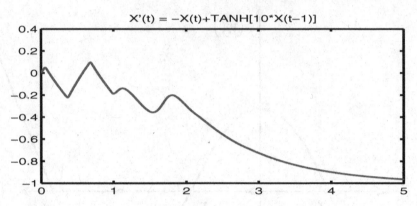

Fig. 5.3 Solution of (5.15) with $\phi(\theta) = \sin(10*\theta)$.

and $x(t,\hat{a}) \nearrow x_0$ whereas $x(t,\hat{b}) \searrow x_0$. Again, by Theorem 5.9, $x(t,\hat{a}) \leq x(t,\phi) \leq x(t,\hat{b})$ if $\phi(s) \in [a,b]$, $s \in [-r,0]$ and consequently $x(t,\phi) \to x_0$.

If $(x-x_0)f(x,x) > 0$, $x \in [a,b], x \neq x_0$ then x_0 is the only equilibrium in $[a,b]$. If $x_0 < c \leq b$ then $f(c,c) > 0$ so, by Proposition 5.10, $t \to x(t,\hat{c})$ is nondecreasing and must leave $[a,b]$ because there is no equilibrium in $(c,b]$. If $\phi(s) \in (x_0,b]$, $s \in [-r,0]$, then $x(t,\phi) \geq x(t,\hat{c})$, $t \geq 0$ for suitable $c \in (x_0, \min \phi)$ by Theorem 5.9. The other case is handled similarly. □

As an example, consider the simple model of a self-excited neuron with delayed excitation given by

$$x'(t) = -x(t) + \tanh(kx(t-1)) \tag{5.15}$$

where $x(t)$ encodes the neuron's "activity level" and $k > 0$. The unit delay, reflecting a scaling of the time variable, represents the transmission time between output $x(t)$ and input.

We do not restrict the sign of x but note that nonnegative (nonpositive) initial data give rise to nonnegative (nonpositive) solutions. Indeed, if $x(t)$ is a solution, then so is $-x(t)$.

Proposition 5.12 *If $0 < k \leq 1$, then $x = 0$ is the only equilibrium and $x(t,\phi) \to 0$ for every initial data.*

If $k > 1$, then $x = 0$ is unstable and there are two additional stable equilibria $x = \pm u$ where $u > 0$ is the unique fixed point of $h(u) = \tanh(ku)$. If $\phi(s) > 0$, $s \in [-1,0]$, then $x(t,\phi) \to u$; if $\phi(s) < 0$, $s \in [-1,0]$, then $x(t,\phi) \to -u$.

Proof. The assertions concerning the equilibria are elementary, using the fact that $h'(0) = k$. The other assertions follow from Corollary 5.11. For example, when $k > 1$ and $\phi(s) > 0$, $s \in [-1,0]$, we can find a,b such that $0 < a < \phi(s), u < b$ with $f(a,a) > 0 > f(b,b)$, implying that $x(t,\hat{a}) \leq x(t,\phi) \leq x(t,\hat{b})$, $t \geq 0$. Because $x(t,\hat{a}) \nearrow u$ and $x(t,\hat{b}) \searrow u$, the conclusion follows. □

As we have seen, there is a strong tendency for solutions of monotone semiflows to converge to equilibrium. Indeed, one can prove that the generic orbit converges to equilibrium and that attracting periodic orbits cannot exist [44, 70, 71]. Furthermore, systems of delay differential equations can generate monotone semiflows [70, 71].

5.5 Delayed Logistic Equation

In 1948, G. Hutchinson ([46]) introduced the delayed logistic equation

$$n'(t) = a[1 - n(t - T)/K]n(t)$$

to model a single population whose percapita rate of growth at time t.

$$n'(t)/n(t) = a[1 - n(t - T)/K]$$

depends on the population size T times units in the past. This would be reasonable for a population that depends on a resource whose density at time t depends on the size of the population feeding on it at time $t - T$ because it takes time T for the resource to recover. If we let $N(t) = n(t)/K$ and rescale time, then we get the discrete-delay logistic equation

$$N'(t) = N(t)[1 - N(t - r)], t \geq 0. \tag{5.16}$$

The right hand side is clearly completely continuous. We are primarily interested in nonnegative solutions. Arguments similar to those used for (3.4) establish that for each $\phi \in C$ with $\phi \geq 0$, there exists a unique nonnegative solution $N(t, \phi)$ defined for all $t \geq 0$. The question is whether $O_+(\phi) = \{N_t(\phi) : t \geq 0\}$ is bounded.

Proposition 5.13 *Every orbit of* (5.16) *with* $\phi \geq 0$ *is bounded. In fact, for each such* ϕ*, there exists* $T > 0$ *such that*

$$0 \leq N(t, \phi) \leq e^r, t > T$$

Proof. There are three cases. If $N'(t) \geq 0$ for all large t, say all $t > t_0$, then $0 \leq N(t) \leq 1$ for $t > t_0 - r$ so we are done. In this case, $N(t) \nearrow 1$, $t \to \infty$ by Proposition 5.7. If $N'(t) \leq 0$ for all $t > t_0$, then $N(t) \geq 1$ for $t > t_0 - r$ but N is decreasing so $N(t) \searrow 1$ as $t \to \infty$ by Proposition 5.7. We are done in this case. Therefore we can assume that no matter how large is t_1, there is a $t_2 > t_1$ and $t_3 > t_1$ such that $N'(t_2) > 0$ and $N'(t_3) < 0$. Of course, this means that $0 \leq N(t_2 - r) < 1$ and $N(t_3 - r) > 1$. Thus N oscillates about $N = 1$. If $t_0 \geq r$ and t_0 is a local maximum of $N(t)$, then $N'(t_0) = 0$ and $N(t_0 - r) = 1$. Therefore, by a now familiar argument, treating (5.16) as a linear equation

$$N(t_0) = N(t_0 - r) \exp(\int_{t_0 - r}^{t_0} [1 - N(s - r)]ds) \leq e^r$$

Therefore e^r is an upper bound for the maximum of N on any interval $[a,b]$ where $N(t) > 1$ on (a,b) and $N(a) = N(b) = 1$ provided $a \geq r$. ☐

Actually, our proof showed that there are potentially three kinds of (nontrivial) solutions of (5.16):

- Solutions that are eventually monotone nondecreasing, less than or equal to one, which converge to the equilibrium $\hat{1}$ from below
- Solutions that are eventually nonincreasing, greater than or equal to one, that converge to one from above
- Solutions that repeatedly oscillate above and below one as $t \to \infty$

We expect that the trivial equilibrium 0 is unstable because the linearization of (5.16) about this solution gives the equation $N' = N$. The following shows that $\omega(\phi) = \{\hat{0}\}$ only in trivial cases.

Proposition 5.14 *Let $\phi \geq 0$. Then $\omega(\phi) = \{\hat{0}\}$ if and only if $\phi(0) = 0$. Indeed, $\phi(0) = 0$ implies that $N(t, \phi) = 0$ for all $t \geq 0$.*

Proof. $N(t) = \phi(0) \exp(\int_0^t [1 - N(s-r)] dr)$ so either $\phi(0) = 0$ and $N(t, \phi) = 0$ for all $t \geq 0$ or $\phi(0) > 0$ and $N(t, \phi) > 0$ for all $t > 0$. In the former case, $\omega(\phi) = \{\hat{0}\}$. Suppose that $\phi(0) > 0$, yet $N(t, \phi) \to 0$ as $t \to \infty$. Then, for some $t_0 > 0$, $N(t) < 1/2$ for $t \geq t_0$ so $N'(t) > \frac{1}{2}N(t)$, $t \geq t_0 + r$, implying that $N(t) \geq N(t_0 + r)e^{t/2} \to \infty$, a contradiction. ☐

If we change variables, putting $u = N - 1$, then (5.16) becomes the famous Wright's equation (see [48, 80])

$$u'(t) = -u(t - r)[1 + u(t)] \tag{5.17}$$

The steady-state $N = 1$ is now $u = 0$ and we expect that if we drop the nonlinearity $u(t)u(t - r)$ in this equation, then the linear equation

$$v'(t) = -v(t - r)$$

will determine the stability of our $N = 1$ steady state. The reader will recall that in Proposition 2.1 and its corollary we provided evidence to the effect that if $r < \pi/2$ then $v = 0$ is asymptotically stable and if $r > \pi/2$, then $v = 0$ is unstable. Therefore, we expect to be able to show that $N = 1$ is asymptotically stable in the case where $r < \pi/2$ and unstable when $r > \pi/2$. Wright [80] proved the following.

Theorem 5.15 *(see [48]) If $r \leq 3/2$, then $N(t, \phi) \to 1$, $t \to \infty$ for all solutions of (5.16) satisfying $\phi(0) > 0$.*

Note that $3/2 = 1.5 < 1.57 \cdots = \pi/2$. Wright's conjecture, still unsolved, is that Theorem 5.15 holds with $\pi/2$ instead of $3/2$. There exist nonconstant periodic solutions of (5.16) when $r > \pi/2$.

Let's first get a positive lower bound for oscillating solutions.

Lemma 5.3. *If $N(t)$ repeatedly oscillates above and below one as $t \to \infty$, then $N(t) \geq \exp(-r(e^r - 1)) > 0$ for all large t.*

Proof. $N(t) \leq e^r$ for all $t > T$ by Proposition 5.13. If $N(t) < 1$ and (a, b) with $N(a) = N(b) = 1$, $a > T + r$, and if N reaches its minimum on (a, b) at t^*, then $N'(t^*) = 0$ so $N(t^* - r) = 1$. It follows that $t^* - r \leq a$ and integrating (5.16) from a to t^* gives

$$N(t^*) = \exp\left(\int_a^{t^*} [1 - N(s - r)] ds \right)$$

$$\geq \exp\left(\int_a^{t^*} [1 - e^r] ds \right)$$

$$= \exp(-(t^* - a)[e^r - 1])$$

$$\geq \exp(-r[e^r - 1])$$

$\square$

It follows from Proposition 5.13 and Lemma 5.3 that oscillatory solutions satisfy

$$0 < \exp(-r(e^r - 1)) \leq N(t) \leq e^r$$

for all large t.

Lemma 5.4. *Let $N(t) > 0$ repeatedly oscillate above and below one as $t \to \infty$ and suppose there exists $0 < A < 1 < B$ and $T > 0$ such that*

$$A \leq N(t) \leq B, t \geq T$$

Then there exists $S > T$ such that

$$(g \circ g)(A) \leq N(t) \leq (g \circ g)(B), t > S$$

where $g(u) := e^{r(1-u)}$.

Proof. The same arguments used in the proof of Proposition 5.13 and Lemma 5.3 tell us that because $1 - B \leq 1 - N(s - r) \leq 1 - A$ holds for large s, we have

$$g(B) \leq N(t) \leq g(A)$$

for large t. Applying this once again, we get the desired result. $\square$

We can show that $N(t) \to 1$, $t \to \infty$ if we can show that for all $x_0 > 0$, the iteration

$$x_{n+1} = g(x_n), n \geq 0$$

satisfies $x_n \to 1$, $n \to \infty$.

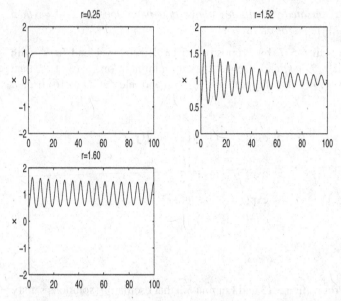

Fig. 5.4 Simulations of (5.16) for different values of r; $N_0 = 0.5$.

Proof. of Theorem 5.15 *when* $r < 1$. We only need to show that $x_n \to 1$, $n \to \infty$. We use Theorem 9.6 of Thieme's book [75]. That theorem says that if $g : (0, \infty) \to (0, \infty)$ is continuous, has a unique fixed point $x^* > 0$, is bounded on $(0, x^*)$, and for some $x_1 < x^* < x_2$, $g(x_1) > x_1$, $g(x_2) < x_2$, and finally, $g \circ g$ has no fixed point other than x^*, then $x_n \to 1$, $n \to \infty$. All requirements of this result are trivial for our g with $x^* = 1$ except the last one: $g(g(u)) = u$ if and only if $g^{-1}(u) = g(u)$ if and only if $1 = F(u) \equiv g(u) + \ln(u)/r$. We show that $F'(u) > 0$ for all u if $r < 1$. If $ruF'(u) = 0$ then $r^2 u g(u) = 1$ but the maximum of the left-hand side occurs at $u = 1/r$ at which it takes the value $re^{r-1} < 1$ inasmuch as $r < 1$ so we conclude that $F' > 0$ and F is injective. We are done!   □

5.6 Delayed Microbial Growth Model

Consider the system (1.17) which, after scaling time and the dependent variables, we may write as

$$S'(t) = 1 - S(t) - f(S(t))x(t) \qquad (5.18)$$
$$x'(t) = \exp(-r)f(S(t-r))x(t-r) - x(t)$$

where

$$f(S) = \frac{mS}{a+S}$$

Theorem 3.4 implies that solutions of (5.18) corresponding to nonnegative initial data (x_0, S_0) are nonnegative on their maximal interval of existence.

Define V by

$$V(t) = x(t) + e^{-r}S(t-r), t \geq r$$

and observe that

$$V'(t) = e^{-r} - V(t)$$

Consequently

$$V(t) \to e^{-r}, t \to \infty \tag{5.19}$$

This bound ensures that nonnegative solutions are defined for all $t \geq 0$ and that they are bounded. Therefore (5.18) generates a semiflow Φ on the positively invariant set $C_+ = C([-r,0], \mathbb{R}_+^2)$. Each orbit has compact closure in C_+ and therefore its omega limit set is nonempty according to Corollary 5.6.

In Chapter 4, Section 4, we found the equilibria for (5.18) and determined their stability properties. It was shown that if $f(1) < e^r$, then the washout equilibrium $(S,x) = (1,0)$ is asymptotically stable. If $f(1) > e^r$, then in addition to the washout equilibrium there is a survival equilibrium $(\bar{S}, \bar{x})$ where $f(\bar{S}) = e^r$ and $\bar{x} = (1-\bar{S})e^{-r}$. In this case, the washout equilibrium is unstable and the survival equilibrium is asymptotically stable.

The following result was first proved by Ellermeyer [28] by the fluctuation lemma. We give a different proof.

Theorem 5.16 *If $f(1) < e^r$, then every solution converges to the washout state. If $f(1) > e^r$, then every solution with $x(0) > 0$ converges to the survival state.*

Proof. Suppose that $f(1) < e^r$. By Exercise 5.14, we may assume that $S(t) < 1$ for large t, so there exist $B \in (0,1)$ and $T > 0$ such that $x'(t) \leq Bx(t-r) - x(t)$, $t \geq T$. We may compare solutions of this differential inequality with solutions of the corresponding differential equation $y'(t) = By(t-r) - y(t)$ satisfying the same initial condition $y_T = x_T$. In fact, $x(t) \leq y(t)$, $t \geq T$ because $z(t) = y(t) - x(t)$ satisfies $z'(t) \geq -z(t) + Bz(t-r)$, $z_T = 0$ and an application of the method of steps establishes that $z(t) \geq 0$, $t \geq T$. By Theorem 4.7(b), with $A = -1$ and B as above, we see that $y(t) \to 0$ at an exponential rate. It follows that $x(t) \to 0$. It is now easy to see from (5.19) that $S(t) \to 1$ and we are done.

Now suppose that $f(1) > e^r$ and suppose $x(0) > 0$. We first show that $x(t)$ cannot converge to zero. Because $x(0) > 0$ and $x' \geq -x$, it follows that $x(t) > 0$, $t \geq 0$. Arguing by contradiction, if $x(t) \to 0$ then $S(t) \to 1$ by Exercise 5.15 and because $e^{-r}f(1) > 1$, there would exist $B > 1$ and $T > r$ such that $x'(t) \geq -x(t) + Bx(t-r)$, $t \geq T$. As in the previous paragraph, we find that $x(t) \geq y(t)$, $t \geq T$ where $y'(t) = -y(t) + By(t-r)$ and $y_T = x_T$. Moreover, $x_T > 0$ inasmuch as $x(t) > 0$, $t \geq r$. The characteristic equation for the linear equation satisfied by y, $\lambda = -1 + Be^{-\lambda r}$, has a positive root λ_0. Let $e(t) = e^{\lambda_0 t}$. There exists $c > 0$ such that $x_T > ce_T$. Then $y(t) \geq z(t)$, $t \geq T$ where $z'(t) = -z(t) + Bz(t-r)$ and $z_T = ce_T$ by our comparison argument again. But $z(t) = ce^{\lambda_0 t}$ and this provides a contradiction because $x(t) \geq z(t)$ and $x(t) \to 0$ by assumption. We conclude that $x^\infty = \limsup_{t \to \infty} x(t) > 0$.

Let $t_n \to \infty$ be such that $x(t_n) \to x^\infty > 0$. Then $\{(S_{t_n}, x_{t_n})\}_n$ has a subsequence converging to an omega limit point $(\phi_1^0, \phi_2^0) \in C_+$ with $\phi_2(0) = x^\infty > 0$. Denote by $(s(t), X(t))$ the solution of (5.18) satisfying $(s_0, X_0) = (\phi_1^0, \phi_2^0)$. By the invariance of ω, it follows that $(X_t, s_t) \in \omega$, $t \geq 0$. As $\phi_2^0 \neq 0$ we have $X(t) > 0$, $t \geq 0$. According to (5.19), we must have $1 = e^r X(t) + s(t - r)$, $t \geq 0$ so $X(t)$ satisfies the scalar. equation

$$X'(t) = e^{-r} f(1 - e^r X(t)) X(t-r) - X(t) \tag{5.20}$$

and $0 < X(t) \leq e^{-r}$ for $t \geq 0$.

If we can show that $X(t) \to \bar{x}$, then it follows that $(s(t), X(t)) \to (\bar{S}, \bar{x})$ and therefore that the asymptotically stable equilibrium $(\hat{\bar{S}}, \hat{\bar{x}})$ belongs to ω. By Exercise 5.10, it follows that $\omega = \{(\hat{\bar{S}}, \hat{\bar{x}})\}$, where we regard $(\hat{\bar{S}}, \hat{\bar{x}})$ as constant functions in C_+. We find functions $u(t)$ and $v(t)$ satisfying $u(t) \leq X(t) \leq v(t)$ and $u(t), v(t) \to \bar{x}$.

Let $u(t)$ satisfy $u'(t) = e^{-r} f(1 - e^r u(t)) u(t-r) - u(t)$, $u_0 = \hat{\varepsilon}$ where $\bar{x} > \varepsilon > 0$ is chosen so small that $e^{-r} f(1 - \varepsilon e^r) > 1$ and $\hat{\varepsilon} \leq X_0$. As $u(t-r) = \varepsilon$, $0 \leq t \leq r$, elementary calculations show that $u(t)$ is increasing on $[0, r]$ so $u_0 \leq u_t$, $0 \leq t \leq r$. By Theorem 5.9, $u_t \leq X_t$ and $u_t \leq \hat{\bar{x}}$ for $0 \leq t \leq r$. By the semiflow property and Theorem 5.9, $u_0 \leq u_t$ implies $u_s \leq u_{t+s}$ for $s > 0$. Hence $u_s \leq u_t \leq \hat{\bar{x}}$ whenever $0 < s < t$ and clearly $u_t \to \hat{\bar{x}}$ as $t \to \infty$.

Let $v(t)$ satisfy $v'(t) = e^{-r} f(1 - e^r v(t)) v(t-r) - v(t)$, $v_0(s) = e^{-r}$. Note that $v'(0) < 0$ and, using the method of steps, one can show that v is strictly decreasing and positive on $[0, r]$. Indeed, because $\hat{\bar{x}} \leq v_0$, we have $\hat{\bar{x}} \leq v_t$ for $t > 0$ by Theorem 5.9. It follows that $\hat{\bar{x}} \leq v_t \leq v_0$ for $t \in [0, r]$, so by the semigroup property and Theorem 5.9, we may conclude that $\hat{\bar{x}} \leq v_t \leq v_s$ whenever $0 < s < t$. Because $X_0 \leq v_0$, we have $X_t \leq v_t$, again by Theorem 5.9. Clearly, $v_t \searrow \hat{\bar{x}}$ as $t \to \infty$.

We have shown that $u_t \leq X_t \leq v_t$, $t \geq 0$ and $u(t), w(t) \to \bar{x}$. It follows that $X(t) \to \bar{x}$. $\square$

Figure 5.5 plots the equilibrium as a function of the delay r. For delay $r > 1$, only the washout equilibrium exists but as r decreases below one, the survival equilibrium appears. Figure 5.6 depicts simulations of (5.18) for $r < 1$ and for $r > 1$.

5.7 Liapunov Functions

Liapunov functions can be used to determine local stability, asymptotic stability, and global stability of equilibria for delay differential equations just as for ordinary differential equations. However, the problem of how to find an appropriate Liapunov function for delay differential equations is even more challenging than for ordinary differential equations. We do not attempt to survey Liapunov methods here. See Hale and Lunel [41] for a thorough treatment. Our goal here is to show that the LaSalle invariance principle can be used to establish global stability.

Consider the system

$$x'(t) = f(x_t)$$

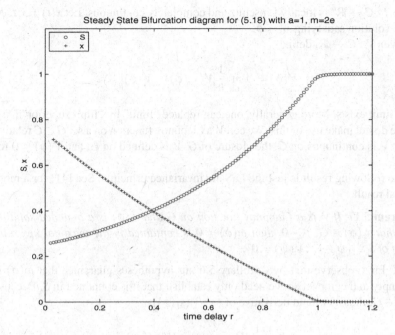

Fig. 5.5 With $m = 2e$ and $a = 1$, $r = 1$ is the bifurcation value.

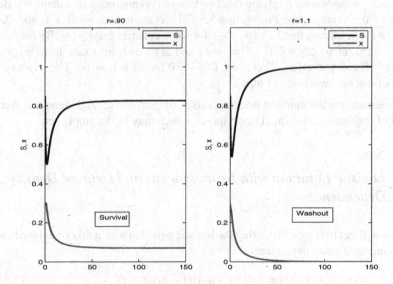

Fig. 5.6 Left: survival at $r = 0.9$; Right: washout at $r = 1.1$.

where $f : C \to \mathbb{R}^n$ is locally Lipschitz and completely continuous. Let $x(t) = x(t, \phi)$ be the solution satisfying $x_0 = \phi$.

Given $V : C \to \mathbb{R}$, define

$$\dot{V}(\phi) = \lim_{h \searrow 0+} \frac{1}{h}[V(x_h(\phi)) - V(\phi)]$$

if the limit exists. More generally, one can replace "limit" by "limit superior" [41], but we do not make use of this. We call V a Liapunov function on a set $G \subset C$ relative to f if V is continuous on $\bar{G}$, the closure of G, $\dot{V}$ is defined on G, and $\dot{V}(\phi) \leq 0$ for $\phi \in G$.

The following result is just the LaSalle invariance principle. See [41] for a more general result.

Theorem 5.17 *If V is a Liapunov function on G and $x_t(\phi)$ is a bounded solution such that $x_t(\phi) \in G$, $t \geq 0$, then $\omega(\phi) \neq \emptyset$ is contained in the maximal invariant subset of $S \equiv \{\psi \in \bar{G} : \dot{V}(\psi) = 0\}$.*

Proof. First, observe that by Corollary 5.6 our hypotheses guarantee that $\omega(\phi)$ is nonempty and compact so we need only establish that it is contained in S. Because $x_t(\phi) \in G$, the right-hand derivative of $t \to V(x_t(\phi))$

$$\lim_{h \searrow 0+} \frac{1}{h}[V(x_{t+h}(\phi)) - V(x_t(\phi))]$$

exists and is nonpositive, implying that $t \to V(x_t(\phi))$ is nonincreasing on $[0, \infty)$. Because $\overline{O_+(\phi)}$ is compact by Proposition 5.5 and V is continuous on $\bar{G}$, V is bounded on $\overline{O_+(\phi)}$ so it follows that $V(x_t(\phi)) \searrow c$ for some constant c as $t \to \infty$. By continuity of V on $\bar{G}$, $\omega(\phi) \subset \{\psi \in \bar{G} : V(\psi) = c\}$ and as $\omega(\phi)$ is invariant, $V(x_t(\psi)) = c$ for all $t \geq 0$ and $\psi \in \omega(\phi)$. Therefore, $\dot{V}(\psi) = 0$ for all $\psi \in \omega(\phi)$. The result now follows from the invariance of $\omega(\phi)$. $\square$

We remark that the union of invariant sets is an invariant set and therefore, every subset of C contains a maximal invariant set, which may be the empty set.

5.7.1 Logistic Equation with Instantaneous and Delayed Density Dependence

We follow Ruan [63] by considering the logistic equation with a mix of delayed and instantaneous density dependence:

$$x'(t) = rx(t)[1 - a_1 x(t) - a_2 x(t - \tau)] \tag{5.21}$$

Parameters r, a_i, τ are positive. The positive equilibrium

$$x^* = \frac{1}{a_1 + a_2}$$

is the focus of our study. We show that it attracts all positive solutions in the case that instantaneous density dependence dominates delayed density dependence:

$$a_1 > a_2 \tag{5.22}$$

First, observe that nonnegative initial data give rise to nonnegative solutions. These solutions are bounded because any nonnegative solution satisfies:

$$x'(t) \leq rx(t)[1 - a_1 x(t)]$$

and therefore $x(t) \leq u(t)$ where $u(t)$ satisfies $u' = ru(1 - a_1 u)$, $u(0) = x(0)$. It follows that nonnegative solutions are continuable to $[0, \infty)$. Lemma 5.14 applies to (5.21) so either $x(t) = 0$ for all $t \geq 0$ or $x(t) > 0$ for all $t \geq 0$. Hereafter, we consider only solutions satisfying $x(0) > 0$.

For future use, we rewrite (5.21) as

$$x'(t) = rx(t)[-a_1(x(t) - x^*) - a_2(x(t - \tau) - x^*)]$$

Let $G = \{\phi \in C : \phi \geq 0, \phi(0) > 0\}$ and define $V : G \to \mathbb{R}$ by

$$V(\phi) = \phi(0) - x^* - x^* \log(\phi(0)/x^*) + \eta \int_{-\tau}^{0} (\phi(s) - x^*)^2 ds \tag{5.23}$$

where $\eta > 0$ is to be determined. In Exercise 5.19, the reader is asked to establish the following facts: (1) V is continuous on G and becomes infinite at a boundary point where $\phi(0) = 0$, and (2) it is positive definite with respect to $X^* \in C$, the function identically equal to x^*, in the sense that

$$V(\phi) > 0 = V(X^*), \phi \in G, \phi \neq X^*.$$

If $x(t, \phi)$ is a solution of (5.21) with $\phi \in G$, then $x_t(\phi) \in G$ for $t \geq 0$ and

$$V(x_t) = x(t) - x^* - x^* \log(x(t)/x^*) + \eta \int_{-\tau}^{0} (x(t+s) - x^*)^2 ds$$

$$= x(t) - x^* - x^* \log(x(t)/x^*) + \eta \int_{t-\tau}^{t} (x(s) - x^*)^2 ds$$

Considered as a function of $t \in [0, \infty)$, it is differentiable and

$$\frac{d}{dt}V(x_t) = \frac{x(t)-x^*}{x(t)}x'(t) + \eta[(x(t)-x^*)^2 - (x(t-\tau)-x^*)^2]$$

$$= \frac{x(t)-x^*}{x(t)}rx(t)[-a_1(x(t)-x^*)-a_2(x(t-\tau)-x^*)]$$

$$+\eta[(x(t)-x^*)^2 - (x(t-\tau)-x^*)^2]$$

$$= -ra_1(x(t)-x^*)^2 - ra_2(x(t)-x^*)(x(t-\tau)-x^*)$$

$$+\eta[(x(t)-x^*)^2 - (x(t-\tau)-x^*)^2]$$

$$= -(ra_1-\eta)(x(t)-x^*)^2 - ra_2(x(t)-x^*)(x(t-\tau)-x^*)$$

$$-\eta(x(t-\tau)-x^*)^2$$

If we take $\eta = ra_1/2$, then

$$\frac{d}{dt}V(x_t) = -(ra_1/2)(x(t)-x^*)^2 - ra_2(x(t)-x^*)(x(t-\tau)-x^*)$$

$$-(ra_1/2)(x(t-\tau)-x^*)^2$$

$$= W(x(t)-x^*,x(t-\tau)-x^*)$$

where W is the quadratic form:

$$W(u,v) = -(ra_1/2)u^2 - ra_2uv - (ra_1/2)v^2$$

$$= -\frac{r}{2}\begin{pmatrix} a_1 & a_2 \\ a_2 & a_1 \end{pmatrix}\begin{pmatrix} u \\ v \end{pmatrix} \cdot \begin{pmatrix} u \\ v \end{pmatrix}$$

The symmetric matrix has eigenvalues $a_1 \pm a_2$, both positive by (5.22), and therefore $W(u,v) < 0$ for $(u,v) \neq (0,0)$. In summary,

$$\frac{d}{dt}V(x_t) = W(x(t)-x^*,x(t-\tau)-x^*) \leq 0, t \geq 0, \phi \in G \qquad (5.24)$$

This leads to the following result, adapted from Ruan [63].

Theorem 5.18 *If (5.22) holds and $\phi \in G$, then $\omega(\phi) = X^*$ and $x(t,\phi) \to x^*$ as $t \to \infty$.*

Proof. By (8.15), $t \to V(x_t(\phi))$ is decreasing on $[0,\infty)$. Assuming that $\phi \neq X^*$, then $\alpha = V(\phi) > 0$. Define $\tilde{G} = \{\psi \in G : V(\psi) \leq \alpha\}$. Then $\tilde{G}$ is closed, positively invariant, and contains $x_t(\phi)$ for all $t \geq 0$. Our calculations above, with $\phi = \psi \in \tilde{G}$ and $t = 0$, show that $\dot{V}(\psi) = W(\psi(0)-x^*,\psi(-\tau)-x^*) \leq 0$. Furthermore, we showed that every orbit of (5.21) is bounded. By Theorem 5.17, $\omega(\phi)$ is a nonempty compact subset of the maximal invariant subset of $S = \{\psi \in \tilde{G} : \dot{V}(\psi) = 0\} = \{\psi \in \tilde{G} : \psi(0) = \psi(-\tau) = x^*\}$. But any invariant subset A of S must have $A = \{X^*\}$ because for each $\psi \in A$ the corresponding solution $x(t)$ satisfies $x_t \in A \subset S$ for $t \geq 0$ and consequently $x_t(-\tau) = x(t-\tau) = x^*$. For $0 \leq t \leq \tau$, this implies that $\psi = X^*$. $\square$

Exercises

Exercise 5.1. Show that the nonautonomous ODE $x' = f(t,x)$ generates a dynamical system by verifying (5.2), assuming that solutions of the initial value problem are unique and extend to all of $\mathbb{R}$. Hint: This just requires uniqueness of solutions of the initial-value problem.

Exercise 5.2. Let $\sigma : I \to X$ and $v : J \to X$ be two solutions of a semidynamical system. If $s \in I \cap J$ and $\sigma(s) = v(s)$ show that $\sigma(t) = v(t)$ for all $t > s$ belonging to $I \cap J$.

Exercise 5.3. Verify (5.4).

Exercise 5.4. Show that the "autonomous" ODE system $x' = f(x)$ generates an autonomous dynamical system assuming unique solutions of the initial-value problem can be extended to all of $\mathbb{R}$.

Exercise 5.5. Show that $\sigma : I \to X$ is a solution corresponding to an autonomous dynamical system Φ provided $t, s \in I$ and $t \geq s$ implies $\sigma(t) = \Phi(t - s, \sigma(s))$.

Exercise 5.6. Let A be an invariant set and $a \in A$. Show that there exists at least one solution $\sigma : \mathbb{R} \to A$ satisfying $\sigma(0) = a$. Hint: $\sigma(t) = \Phi(t,a)$, $t \geq 0$. Because there exists $a' \in A$ such that $\Phi(1, a') = a$, extend $\sigma : [-1, \infty) \to X$ by $\sigma(t) = \Phi(1 + t, a')$. Be sure to show that this new definition of σ agrees with the previous one on $[0, \infty)$. Now use mathematical induction. Do you see why this solution may not be unique?

Exercise 5.7. Let $a, b \in \mathbb{R}$ satisfy $a < b$. Show that if $x : [a - r, b) \to \mathbb{R}^n$ is a solution of (5.7) and $c \in \mathbb{R}$, then $x(t + c)$ is a solution on $[a - c - r, b - c)$.

Exercise 5.8. Determine the equilibria of equations (3.7), (3.8), (3.9), and (3.19).

Exercise 5.9. Let $f(\phi) = F(\phi(0), \phi(-r))$ for some continuous function $F : \mathbb{R}^{2n} \to \mathbb{R}^n$ in which case (5.7) becomes $x'(t) = F(x(t), x(t - r))$. Show that f is completely continuous.

Exercise 5.10. Suppose that a stable equilibrium e belongs to the omega limit set $\omega(\phi)$ of the orbit through ϕ. Show that $\omega(\phi) = \{e\}$.

Exercise 5.11. Assume that f is continuously differentiable. Let x_0 be an equilibrium of (5.11) and (5.13). Let $A = f_x(x_0, x_0)$ and $B = f_y(x_0, x_0)$. Then $B \geq 0$. If $A + B > 0$, then x_0 is unstable for (5.13). Show that it is unstable for (5.11). If $A + B < 0$, then x_0 is asymptotically stable for (5.13). Show that it is asymptotically stable for (5.11). Use Theorem 4.7

Exercise 5.12. Show that if $\phi \in C$ satisfies $\phi \geq 0$ then the solution $N(t, \phi)$ of Nicholson's blowfly equation

$$N'(t) = -\delta N(t) + pN(t - r)\exp(-qN(t - r)) \qquad (5.25)$$

satisfies $N(t, \Phi) \geq 0$, $t \geq 0$ and $O_+(\phi)$ is bounded. $\delta, p, q, r > 0$.

Exercise 5.13. Use the MATLAB software package to corroborate the assertions made concerning (5.16).

Exercise 5.14. For (5.18), use the differential inequality $S'(t) \leq 1 - S(t)$ to show that either $S(t) > 1$ for all t, in which case $S(t) \searrow 1$, $t \to \infty$, or $S(t) < 1$ for all large t.

Exercise 5.15. For system (5.18), if $x(t) \to 0$ show that $S(t) \to 1$. Hint: Use differential inequalities related to the equation for S.

Exercise 5.16. The proof of Theorem 5.16 includes an analysis of the monotone limiting equation (5.20) on the positive invariant set $\{\phi \in C : \hat{0} \leq \phi \leq \hat{x}\}$. It showed that all nonzero solutions converge to $\hat{x}$ if $f(1) > e^r$. Show that if $f(1) < e^r$ then all solutions converge to zero.

Exercise 5.17. Show that the Heine–Borel theorem fails to hold for C.

Exercise 5.18. Provide missing details in the proof of Proposition 5.12. Use MATLAB to investigate the behavior of the solution of (5.15) for $k > 1$ corresponding to the initial data $\phi(s) = m(s + 0.5)$, $-1 \leq s \leq 0$ where $m > 0$ is so large that the image of ϕ includes $[-u, u]$ in its interior.

Exercise 5.19. Verify that (5.23) satisfies: (1) V is continuous on G and becomes infinite at a boundary point where $\phi(0) = 0$, and (2) it is positive definite with respect to $X^* \in C$, the function identically equal to x^*, in the sense that

$$V(\phi) > 0 = V(X^*), \phi \in G, \phi \neq X^*.$$

Exercise 5.20. Calculate the characteristic equation associated with the linearization of (5.21) about x^* and show that x^* can become unstable if $a_2 \geq a_1$.

Exercise 5.21. Arino, Wang, and Wolkowicz [4] derive the following model of single-species growth

$$N'(t) = \gamma F(N(t - \tau)) - \mu N(t) - \kappa N^2(t)$$

where

$$F(N) = \frac{\mu N}{\mu e^{\mu \tau} + \kappa (e^{\mu \tau} - 1)N}$$

where $\gamma, \mu, \kappa, \tau > 0$. Notice that $\tau = 0$ results in the undelayed logistic equation and gives a clue to the derivation of the equation above.

1. Show that the equation generates a monotone semiflow on the nonnegative states $C_+ = \{\phi \in C : \phi \geq 0\}$.
2. Find conditions for the existence of a positive equilibrium.
3. Determine the stability of the trivial equilibrium and the positive one.

Exercise 5.22. Theorem 5.18 shows that x^* is globally stable for positive initial data but it does not mean that it is a stable equilibrium. Show that X^* is an asymptotically stable equilibrium. Hint: Use V.

Exercise 5.23. Consider the equation $x'(t) = f(x(t), x(t-r))$ where $f \in C^1$. Let $a < b$ and $G = \{\phi \in C : a \le \phi(\theta) \le b, \theta \in [-r, 0]\}$. Show that G is positively invariant if

$$\phi \in G, \phi(0) = b \Rightarrow f(\phi(0), \phi(-r)) \le 0$$

and the analogous condition when $\phi(0) = a$, with inequality reversed, hold. Hint: Mimic the proof of Proposition 5.10, using Theorem 3.4. See Chapter 5, Remark 2.1 of [71] for a generalization to systems.

Exercise 5.24. Cooke [19] models a vector borne disease by the equation

$$y'(t) = by(t-r)[1 - y(t)] - cy(t)$$

where $b, r > 0$ and $c \ge 0$. $y(t)$ denotes the fraction of human population that is infected with the disease, c is the recovery rate, r is the delay between when a vector (e.g., a mosquito) becomes infected after biting an infected person until the time when its bite can infect a human, and b is the contact rate. Show that $G = \{\phi \in C : 0 \le \phi(\theta) \le 1\}$ is positively invariant as the model demands. If $c > b > 0$, show that the trivial solution is globally attracting. You may either use monotone dynamics or the Lyapunov function

$$V(\phi) = \frac{1}{2c}\phi(0)^2 + \frac{1}{2}\int_{-r}^0 \phi^2(\theta)d\theta$$

Exercise 5.25. For autonomous ODEs, unique solutions of the initial-value problem exist forward and backward in time resulting in a flow map $\Phi : \mathbb{R} \times X \to X$ for which each time t map $x \to \Phi(t, x)$ is one-to-one. This fails for delay differential equations: the semiflow Φ may have the property that the time t map $\phi \to \Phi(t, \phi)$ is not one-to-one. Show that the semiflow generated by (5.16) gives an example by using Proposition 5.14.

Chapter 6
Hopf Bifurcation

Abstract The Hopf bifurcation theorem is one of the most important results for delay differential equations because it is essentially the only method for rigorously establishing the existence of periodic solutions. We begin with an example where the bifurcation can be easily calculated in order to show key features of a Hopf bifurcation. Then the theorem is stated without proof along with a discussion of the stability properties of bifurcating periodic solutions. The remainder of the chapter consists of numerous applications. A complete study of the Hopf bifurcation for the nonlinear negative feedback equation is carried out with particular application to the delayed logistic equation. Hopf bifurcation for second-order delayed negative feedback equations is thoroughly studied, including the inverted pendulum equation with delayed feedback. A gene regulatory network provides another example. The chapter concludes with a statement and elaboration on a beautiful Poincaré-Bendixson theorem due to Mallet-Paret and Sell for a special class of delay differential systems, namely, monotone cyclic feedback systems.

6.1 A Canonical Example

Consider the delay differential equation

$$x'(t) = x(t-1)[-v + x^2(t) + x^2(t-1)] \tag{6.1}$$

where v is a real parameter. Its steady states $x = a$ are given as solutions of

$$0 = a[-v + 2a^2]$$

so $a = 0$ is a steady state for all v and $a = \pm\sqrt{v/2}$ is a steady state for $v > 0$.

Let's focus on the steady-state $x = a = 0$ first. Linearizing about $x = 0$ by dropping the higher order terms $x^2(t) + x^2(t-1)$ from (6.1), we get the linear equation

$$y'(t) = -vy(t-1) \tag{6.2}$$

H. Smith, *An Introduction to Delay Differential Equations with Applications to the Life Sciences*, 87
Texts in Applied Mathematics 57, DOI 10.1007/978-1-4419-7646-8_6,
© Springer Science+Business Media, LLC 2011

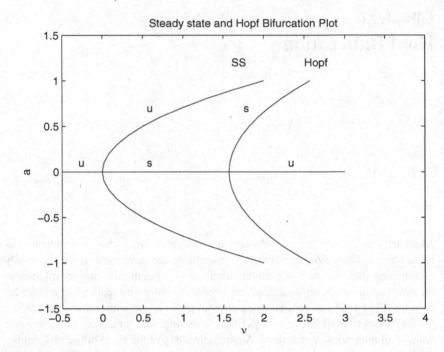

Fig. 6.1 Steady-state and Hopf bifurcation for (6.1).

which was considered in great detail in our first section. Its characteristic equation is the familiar one:

$$\lambda = -ve^{-\lambda}$$

From Proposition 2.1 we know that $y = 0$ is asymptotically stable for $0 < v < \pi/2$ inasmuch as the roots have negative real part. At $v = 0$, $\lambda = 0$ is the only root. At $v = \pi/2$, $\lambda = \pm i\pi/2$ are roots. Therefore $x = 0$ is asymptotically stable for (6.1) when $0 < v < \pi/2$.

We would like to know what happens to solutions of (6.1) as v passes through 0 from positive to negative and we'd like to know what happens as v passes through $\pi/2$. Let's take up the case $a \approx 0$ first. The linear equation (6.2) ought to tell us something. For $v \neq 0$, its only steady state is $y = 0$ but when $v = 0$ every value of y gives a steady state. Thus there is a major change in the structure of the set of steady states for the linear equation (6.2) at $v = 0$. But we saw that the same happens for the nonlinear equation (6.1). Namely, the new branch of steady states

$$a = \pm\sqrt{v/2}, v > 0$$

branches off from the trivial steady state at $v = 0$. This is called a *steady-state bifurcation* and it occurs precisely where the linear equation (6.2) predicted it would, at $v = 0$.

What does the linear equation predict will happen for v near $\pi/2$? At $v = \pi/2$, $y(t) = \sin(\pi t/2)$ is a solution of (6.2), as is $y(t) = \cos(\pi t/2)$. Thus the linear equation is predicting that a periodic solution appears suddenly at $v = \pi/2$. We had better listen to what the linear equation predicts. Let's try

$$x(t) = a\sin(\pi t/2)$$

as a solution of (6.1). Using that $\sin(\pi(t-1)/2) = -\cos(\pi t/2)$, we have, on inserting the above into (6.1) and dividing out the common factor $\cos(\pi t/2)$ in every term:

$$a\frac{\pi}{2} = av - a^3$$

Dividing out a factor of a, we have that $a = \pm\sqrt{v - \pi/2}$ so

$$x(t) = \pm\sqrt{v - \frac{\pi}{2}} \cdot \sin(\pi t/2), v > \frac{\pi}{2}$$

is a periodic solution of (6.1).

6.2 Hopf Bifurcation Theorem

Now let's state the Hopf bifurcation theorem, following [26, 41, 43]. We consider the one-parameter family of delay equations

$$x'(t) = F(x_t, \mu) \tag{6.3}$$

where $F : C \times \mathbb{R} \to \mathbb{R}^n$ is twice continuously differentiable in its arguments and $x = 0$ is a steady state for all values of μ:

$$F(0, \mu) \equiv 0.$$

We may linearize F about $\phi = 0$ as follows

$$F(\phi, \mu) = L(\mu)\phi + f(\phi, \mu)$$

where $L(\mu) : C \to \mathbb{R}^n$ is a bounded linear operator and f is higher order:

$$\lim_{\phi \to 0} \frac{|f(\phi, \mu)|}{\|\phi\|} = 0$$

The characteristic equation associated with L is

$$0 = \det(\lambda I - A(\mu, \lambda)), A_{ij}(\mu) := L(\mu)_i(e_\lambda e_j)$$

The main assumption is on the characteristic roots of this equation.

(H) For $\mu = 0$, the characteristic equation has a pair of simple roots $\pm i\omega_0$ with $\omega_0 \neq 0$ and no other root that is an integer multiple of $i\omega_0$.

By a simple root, we mean a root of order one. If we express the characteristic equation as $h(\mu, \lambda) = 0$, then (H) implies that the partial derivative $h_\lambda(0, i\omega_0) \neq 0$. Therefore, the implicit function theorem implies the existence of a continuously differentiable family of roots $\lambda = \lambda(\mu) = \alpha(\mu) + i\omega(\mu)$ for small μ satisfying $\lambda(0) = i\omega_0$. In particular, $\alpha(0) = 0$ and $\omega(0) = \omega_0$. Our next assumption is that these roots cross the imaginary axis transversally as μ increases through zero. More precisely, we assume that:

$$\alpha'(0) > 0. \tag{6.4}$$

Note that if $\alpha'(0) < 0$, then we can always arrange that (6.4) holds by the change of parameter $v = -\mu$. Thus, the positive sign is simply a normalization that ensures that for $\mu < 0$, the pair of roots has negative real part whereas for $\mu > 0$ it has positive real part.

Our formulation below follows Theorem 2.7 of Chapter X of [26] but uses Lemma 10.2 of Chapter 7 of [41] to reformulate hypothesis $(H\zeta 3)$ of Theorem 2.7. See also Theorem 3.9 of Chapter X of [26] for the assertions regarding direction of bifurcation.

Theorem 6.1 *Let (H) and (6.4) hold. Then there exists $\varepsilon_0 > 0$, real-valued even functions $\mu(\varepsilon)$ and $T(\varepsilon) > 0$ satisfying $\mu(0) = 0$ and $T(0) = 2\pi/\omega_0$, and a nonconstant $T(\varepsilon)$-periodic function $p(t, \varepsilon)$, with all functions being continuously differentiable in ε for $|\varepsilon| < \varepsilon_0$, such that $p(t, \varepsilon)$ is a solution of (6.3) and $p(t, \varepsilon) = \varepsilon q(t, \varepsilon)$ where $q(t, 0)$ is a $2\pi/\omega_0$-periodic solution of $q' = L(0)q$.*

Moreover, there exist $\mu_0, \beta_0, \delta > 0$ such that if (6.3) has a nonconstant periodic solution $x(t)$ of period P for some μ satisfying $|\mu| < \mu_0$ with $\max_t |x(t)| < \beta_0$ and $|P - 2\pi/\omega_0| < \delta$, then $\mu = \mu(\varepsilon)$ and $x(t) = p(t + \theta, \varepsilon)$ for some $|\varepsilon| < \varepsilon_0$ and some θ.

If F is five times continuously differentiable then:

$$\mu(\varepsilon) = \mu_1 \varepsilon^2 + O(\varepsilon^4)$$
$$T(\varepsilon) = \frac{2\pi}{\omega_0}[1 + \tau_1 \varepsilon^2 + O(\varepsilon^4)] \tag{6.5}$$

If all other characteristic roots for $\mu = 0$ have strictly negative real parts except for $\pm i\omega_0$ then $p(t, \varepsilon)$ is asymptotically stable if $\mu_1 > 0$ and unstable if $\mu_1 < 0$.

The case that $\mu_1 > 0$ is called a "supercritical" Hopf bifurcation. If this is the case and if all other characteristic roots for $\mu = 0$ have strictly negative real parts except for $\pm i\omega_0$, then steady state $x = 0$ is asymptotically stable for $\mu < 0$ near zero and unstable for $\mu > 0$ near zero. For $\mu = \mu_1 \varepsilon^2 + O(\varepsilon^4) > 0$ an asymptotically stable periodic solution $p(t, \varepsilon)$ exists. Stability is lost by the steady-state and gained by the periodic solution. Our canonical example (6.1) undergoes a supercritical Hopf bifurcation at $\mu = 0$ where $v = \pi/2 + \mu$. The periodic solution $p(t, \varepsilon) = \varepsilon \sin(\pi t/2)$

corresponds to $\mu = \varepsilon^2$ so $\mu_1 = 1$. According to Theorem 6.1, the bifurcating periodic solution is asymptotically stable.

A quite different qualitative picture emerges when $\mu_1 < 0$, the so-called "sub-critical" Hopf bifurcation. In this case, the unstable periodic solution $p(t, \varepsilon)$ exists when $\mu = \mu_1 \varepsilon^2 + O(\varepsilon^4) < 0$. Although the steady-state $x = 0$ is asymptotically stable for $\mu < 0$, its basin of attraction must be very small when μ is small because the unstable periodic solution of amplitude proportional to $\sqrt{-\mu}$ is nearby. Most disturbingly, when $x = 0$ loses stability for $\mu > 0$, no small bounded solution gains stability. If there is a stable bounded solution for $\mu > 0$, it is "large".

Our stability assertions require elaboration because a nonconstant periodic solution $x(t)$ of (6.3) cannot be asymptotically stable in the usual sense because $y(t) = x(t + \theta)$ is also a periodic solution, and if $\theta \neq 0$ is small, results in $\|x_0 - y_0\|$ being small yet $\|x_t - y_t\| \nrightarrow 0$ as $t \to \infty$. Our stability assertions are in the orbital sense: the periodic solution $x(t)$ is orbitally stable if for every $\varepsilon > 0$ there exists $\delta > 0$ such that if $d(y_0, O) < \delta$, then $d(y_t, O) < \varepsilon$ for all $t \geq 0$ where $O = \{x_t : t \in \mathbb{R}\} \subset C$ and $d(y_0, O) = \inf\{\|y_0 - \phi\| : \phi \in O\}$. It is orbitally asymptotically stable if it is orbitally stable and there exists $b > 0$ such that if $d(y_0, O) < b$ then there exists θ such that $\|y_t - x_{t+\theta}\| \to 0$ as $t \to \infty$.

Methods for computing the expansions (6.5) are described in [43] and [26]. Like their ODE counterparts, they are computationally challenging.

6.3 Delayed Negative Feedback

Consider the canonical equation with single delay

$$x'(t) = -f(x(t - r))$$

where the smooth function f satisfies:

$$f(0) = 0, f'(0) = 1, f''(0) = A, f'''(0) = B$$

The equation exhibits negative feedback in view of the negative sign and in the sense that $xf(x) > 0$, at least for x near zero. In the absence of the delay, $x = 0$ would be an asymptotically stable equilibrium. For general such f, this equation has been extensively studied in the literature. See especially Mallet-Paret [55] and Chapter XVI in [26]. Entire monographs have been devoted to the study of its dynamics.

It is worth pointing out that if f is C^∞, that is, it is differentiable of all orders, then any periodic solution must also be infinitely many times differentiable. This is immediate from the method of steps because a solution gains an extra derivative after every delay interval but it exists on all of $\mathbb{R}$ and repeats itself.

It is convenient to get the parameter r out of the argument of x, so we scale time by $s = t/r$. This results in the equation with fixed delay

$$\dot{x}(s) = -rf(x(s - 1)) \tag{6.6}$$

The linearized equation about $x = 0$ is just

$$\dot{v}(s) = -rv(s - 1) \tag{6.7}$$

with the familiar characteristic equation

$$\lambda + re^{-\lambda} = 0$$

As we know, the roots have negative real part when $0 \le r < \pi/2$; $\lambda = \pm i\pi/2$ are roots at $r = \pi/2$ and all other roots have negative real part so it is natural to set $r = \pi/2 + \mu$ to connect with the notation of the Hopf bifurcation theorem. Writing the characteristic equation as

$$F(\lambda, \mu) := \lambda + (\pi/2 + \mu)e^{-\lambda} = 0 \tag{6.8}$$

we find that

$$F(i\pi/2, 0) = 0, F_\lambda(i\pi/2, 0) = 1 + (\pi/2)i, F_\mu(i\pi/2, 0) = -i$$

so the implicit function theorem implies that we can solve $F = 0$ for $\lambda = \lambda(\mu) = \alpha(\mu) + i\omega(\mu)$ satisfying $\lambda(0) = (\pi/2)i$ and

$$\frac{d\lambda}{d\mu}(0) = -F_\mu/F_\lambda = \frac{\pi/2 + i}{1 + (\pi/2)^2}$$

Consequently,

$$\frac{d\alpha}{d\mu}(0) = \frac{\pi/2}{1 + (\pi/2)^2}. \tag{6.9}$$

When $\mu = 0$ there exists $\delta > 0$ such that the roots of the characteristic equation consist of $\pm i(\pi/2)$ and roots λ satisfying $\Re(\lambda) < -\delta$. See Proposition 2.1. Consequently, by the continuity of the roots with respect to μ, there exists $\mu_0 > 0$ such that for $|\mu| < \mu_0$, the roots of (6.8) consist of $\alpha(\mu) \pm i\omega(\mu)$ and other roots satisfy $\Re(\lambda) < -\delta/2$. Thus the hypotheses of the Hopf bifurcation theorem are satisfied. As a consequence, (6.6) has a nonconstant $T(\varepsilon)$-periodic solution $p(t, \varepsilon)$ corresponding to $r = \pi/2 + \mu(\varepsilon)$ with $\mu(0) = 0$ and $T(0) = 2\pi/(\pi/2) = 4$. Below, we determine whether it exists for $r > \pi/2$ or for $r < \pi/2$. As $\alpha'(0) > 0$, the former implies the periodic solutions are stable and the latter implies they are unstable.

6.3.1 Computation of the Hopf Bifurcation

As the linearized equation (6.7) has periodic solutions $\sin(\pi s/2)$, $\cos(\pi s/2)$ when $r = \pi/2$ we expect these to be leading terms in any periodic solution that bifurcates from the zero solution when $r \approx \pi/2$. We make a further change of the time variable in order to use standard Fourier series formulas in our calculations; in order to make

$\sin(\pi s/2)$, $\cos(\pi s/2)$ the fundamental mode, we set

$$\tau = \pi s/2, R = 2r/\pi$$

The new equation for $X(\tau) = x(s)$ becomes

$$\frac{dX}{d\tau}(\tau) = -Rf\left(X(\tau - \frac{\pi}{2})\right) \tag{6.10}$$

Its linearization

$$\frac{dY}{d\tau}(\tau) = -RY(\tau - \frac{\pi}{2})$$

has the periodic solutions $\sin(\tau)$, $\cos(\tau)$ when $R = 1$. Therefore, we expect small periodic solutions with these as leading terms for $R \approx 1$. The computation of the periodic solution as a power series in a small parameter will require us to find periodic solutions of the periodically forced linear system

$$\frac{dP}{dz}(z) + P(z - \frac{\pi}{2}) = h(z) = h(z + 2\pi)$$

Does this equation have a 2π-periodic solution for a given 2π-periodic forcing function h? Complex Fourier series are easiest to work with here so we use them:

$$h = \sum_{n \in Z} h_n e^{inz}$$

is the Fourier series for h, where

$$h_n = \frac{1}{2\pi} \int_{-\pi}^{\pi} h(z) e^{-inz} dz, n \in \mathbb{Z}$$

It converges to h in the mean square sense.

Both h and the solution P have Fourier series and we can solve for P if we can determine its Fourier coefficients in terms of those of h. The relevant series are:

$$h = \sum_{n \in Z} h_n e^{inz}, P = \sum_{n \in Z} P_n e^{inz}$$

$$P(\bullet - \frac{\pi}{2}) = \sum_{n \in Z} P_n e^{-in\frac{\pi}{2}} e^{inz}$$

$$\frac{dP}{dz} = \sum_{n \in Z} in P_n e^{inz}$$

Inserting these into our equation and equating coefficients of e^{inz} leads to

$$(in + e^{-in\frac{\pi}{2}})P_n = h_n, n \in \mathbb{Z}$$

Consequently, as the term in parentheses vanishes if and only if $n = \pm 1$, we find that there is a 2π-periodic solution P if and only if

$$h_1 = h_{-1} = 0 \tag{6.11}$$

and its Fourier coefficients are given by

$$P_n = \frac{h_n}{in + (-i)^n}, |n| > 1, P_0 = h_0 \tag{6.12}$$

and where P_1 and P_{-1} are arbitrary.

For $k \geq 0$, let

$$H^k = \{h \in L^2(\mathbb{T}) : \sum_{n \in Z} n^{2k} |h_n|^2 < \infty\}$$

These spaces are Hilbert spaces contained in the Hilbert space $H^0 = L^2(\mathbb{T})$ of square integrable functions on the unit circle $\mathbb{T}$. The larger is k, the smoother are the functions in H^k. In the exercises the reader is asked to show that if $h \in H^k$ then $P \in H^{k+1}$.

The reader need not be intimidated by the fancy Hilbert spaces introduced above inasmuch as we only use the above considerations when h is a trigonometric polynomial, that is, when its Fourier series has finitely many terms; then no issues of convergence of Fourier series arise.

6.3.2 Series Expansion of Hopf Solution

We expect that the period of the periodic solution of (6.10) will deviate from 2π. Because it is convenient to work in the space of 2π-periodic functions, we are motivated to introduce an a priori unknown frequency ω:

$$X(\tau) = P(\omega\tau)$$

where P is 2π-periodic and where $\omega \approx 1$ when $R \approx 1$. The equation satisfied by $P(z)$ is

$$\frac{dP}{dz}(z) = -\frac{R}{\omega}f(P(z - \frac{\pi}{2}\omega)) \tag{6.13}$$

As this equation is autonomous, time-translates of solutions are again solutions, so if it has a periodic solution with leading terms involving $\sin(z)$, $\cos(z)$ when $R, \omega \approx 1$, that is, if its real Fourier series is:

$$P(z) = \frac{P_0}{2} + P_1 \cos(z) + Q_1 \sin(z) + \text{higher harmonics}$$

where $P_1^2 + Q_1^2 \neq 0$, then after a time translation, we may assume that:

$$Q_1 = 0$$

Therefore, following (6.5), we seek a solution of the form:

$$P(z) = \varepsilon p_0(z) + \varepsilon^2 p_1(z) + \varepsilon^3 p_2(z) + \cdots$$
$$R = 1 + \varepsilon^2 r_1 + \varepsilon^4 r_2 + \cdots \qquad (6.14)$$
$$\omega = 1 + \varepsilon^2 \omega_1 + \varepsilon^4 \omega_2 + \cdots$$

where p_i are 2π-periodic and

$$p_0(z) = \cos(z), \int_{-\pi}^{\pi} p_i(z)\cos(z)dz = \int_{-\pi}^{\pi} p_i(z)\sin(z)dz = 0, n \geq 1$$

What do we want to learn? We already expect that $P(z) = \varepsilon \cos(z) + \cdots$ and the higher-order corrections $p_i(z)$ are not so interesting. What we really want to know is the value of r_1 because it tells us whether the periodic solutions exist for $R > 1$ or for $R < 1$; the former is a supercritical Hopf bifurcation, the latter is subcritical Hopf bifurcation. It might be nice to know ω_1 but most of the time we don't care about it. Keep in mind in the following calculations that we have only these limited objectives in mind.

Inserting (6.14) into (6.13), the left-hand side is

$$\varepsilon \frac{dp_0}{dz}(z) + \varepsilon^2 \frac{dp_1}{dz}(z) + \varepsilon^3 \frac{dp_2}{dz}(z) + \cdots$$

The right-hand side is more complicated. We begin with some baby steps; note that ω is a function of ε.

$$\frac{R}{\omega} = 1 + \varepsilon^2 (r_1 - \omega_1) + O(\varepsilon^4)$$

$$P(z - \frac{\pi}{2}\omega) = \varepsilon p_0(z - \frac{\pi}{2}\omega) + \varepsilon^2 p_1(z - \frac{\pi}{2}\omega) + \varepsilon^3 p_2(z - \frac{\pi}{2}\omega) + \cdots$$

$$= \varepsilon p_0(z - \frac{\pi}{2}) + \varepsilon^2 p_1(z - \frac{\pi}{2})$$

$$+ \varepsilon^3 \left(p_2(z - \frac{\pi}{2}) - \frac{\pi}{2}\omega_1 \frac{dp_0}{dz}(z - \frac{\pi}{2}) \right) + O(\varepsilon^4)$$

We use

$$zs = z - \frac{\pi}{2}$$

to simplify notation. Inserting the expansion of $P(z - \pi\omega/2)$ into

$$f(x) = x + \frac{A}{2}x^2 + \frac{B}{6}x^3 + \cdots$$

we find that

$$f(P(z - \frac{\pi}{2}\omega)) = \varepsilon p_0(zs) + \varepsilon^2 \left(p_1(zs) + \frac{A}{2}p_0^2(zs) \right)$$

$$+ \varepsilon^3 \left(p_2(zs) - \frac{\pi}{2}\omega_1 \frac{dp_0}{dz}(zs) + Ap_0(zs)p_1(zs) + \frac{B}{6}p_0^3(zs) \right)$$

$$+ O(\varepsilon^4)$$

Now multiply the above by $-R/\omega$ and match like powers of ε on the right and left sides of (6.13) to get

$$\frac{dp_0}{dz}(z) = -p_0(zs)$$

$$\frac{dp_1}{dz}(z) = -p_1(zs) - \frac{A}{2}p_0^2(zs)$$

$$\frac{dp_2}{dz}(z) = -p_2(zs) + \frac{\pi}{2}\omega_1 \frac{dp_0}{dz}(zs) - (r_1 - \omega_1)p_0(zs)$$

$$-Ap_0(zs)p_1(zs) - \frac{B}{6}p_0^3(zs)$$

The equation for p_0 is satisfied by $p_0(z) = \cos(z)$ as we expected. Making liberal use of trig identities, the equation for p_1 is

$$\frac{dp_1}{dz}(z) = -p_1(z - \frac{\pi}{2}) - \frac{A}{4} + \frac{A}{4}\left(\frac{e^{2iz} + e^{-2iz}}{2} \right)$$

As the inhomogeneous term satisfies (B.3), the unique solution (see (B.4)) that is orthogonal to both $\sin(z)$, $\cos(z)$ is

$$p_1(z) = -A\left[\frac{1}{4} + \frac{\cos(2z)}{20} - \frac{\sin(2z)}{10} \right]$$

Ok, now it's going to get messy. We insert p_1 and p_0 into the equation for p_2. Fortunately, we don't have to solve the equation for p_2 inasmuch as we do not care what it is. We want only to determine r_1 and ω_1. The equation for p_2 is

$$\frac{dp_2}{dz}(z) = -p_2(z - \frac{\pi}{2}) + h(z)$$

where h is given by

$$h(z) = \frac{\pi}{2}\omega_1 \cos(z) + (\omega_1 - r_1 + \frac{A^2}{4})\sin(z)$$

$$A^2 \sin(z)\left(\frac{\sin(2z)}{10} - \frac{\cos(2z)}{20} \right) - \frac{B}{6}\sin^3(z)$$

We only need to determine the Fourier $\sin z$ and $\cos(z)$ coefficients of h because these must vanish, by (B.3), in order that there be a 2π-periodic solution and because the vanishing of these coefficients will determine r_1 and ω_1. Using

$$\sin^3(z) = \frac{3}{4}\sin(z) - \frac{1}{4}\sin(3z)$$

$$\sin(z)\cos(2z) = \frac{1}{2}(\sin(3z) - \sin(z))$$

$$\sin(z)\sin(2z) = \frac{1}{2}(\cos(z) - \cos(3z))$$

leads to

$$h(z) = \frac{h_0}{2} + \left(\frac{\pi}{2}\omega_1 + \frac{A^2}{20}\right)\cos(z) + \left(\omega_1 - r_1 + \frac{11A^2}{40} - \frac{B}{8}\right)\sin(z)$$
$$+\text{higher harmonics}$$

For p_2 to be 2π-periodic we must have

$$0 = \frac{\pi}{2}\omega_1 + \frac{A^2}{20}$$

$$0 = \omega_1 - r_1 + \frac{11A^2}{40} - \frac{B}{8}$$

and so

$$\omega_1 = -\frac{A^2}{10\pi} \tag{6.15}$$

$$r_1 = A^2\left(\frac{11\pi - 4}{40\pi}\right) - \frac{B}{8} \tag{6.16}$$

Supercritical Hopf bifurcation occurs if $r_1 > 0$, that is, when

$$A^2\left(\frac{11\pi - 4}{5\pi}\right) > B \tag{6.17}$$

Subcritical bifurcation holds if the reverse inequality holds. Note that $(11\pi - 4)/5\pi \approx 1.95$.

Observe that the direction of bifurcation carries over to the original equation (6.6) that motivated this section.

An approach to determine the stability of a Hopf bifurcation using center manifold techniques can be found in [33].

6.3.3 The Logistic Equation

The logistic equation

$$N'(t) = N(t)[1 - N(t - r)] \tag{6.18}$$

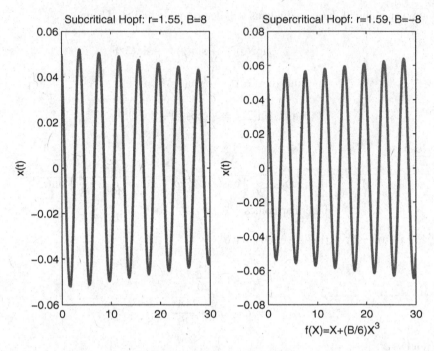

Fig. 6.2 Solution plots for case $f(x) = x + (B/6)x^3$.

has positive equilibrium $N = 1$. The change of variables

$$v = N - 1$$

transforms this equilibrium to the origin. The equation for v is

$$v'(t) = -v(t - r)(1 + v(t))$$

The change of variables

$$x = \log(1 + v)$$

converts the latter to the delayed negative feedback equation

$$x'(t) = -f(x(t - r))$$

where

$$f(x) = e^x - 1$$

Therefore, in the notation of our general delayed feedback equation (6.6), we have

$$A = B = 1$$

This implies by (6.17) that the logistic equation has a supercritical Hopf bifurcation at $r = \pi/2$ which is asymptotically stable.

6.4 A Second-Order Delayed Feedback System

Suppose we control the external forcing $F(t)$ to the damped oscillator equation

$$x'' + bx' + ax = F(t), a, b > 0$$

We'd like to control $x(t)$ in order to "stabilize" the system near $x = 0$. A common engineering approach is to observe $x(t)$ and, depending on its deviation from $x = 0$, apply a suitable forcing $F(t)$. In other words, we want to use *state feedback*

$$F(t) = F(x(t))$$

However, there is a delay in our ability to implement the forcing and thus it is more reasonable to consider

$$F(t) = F(x(t - r))$$

We assume that the negative feedback condition for F

$$xF(x) < 0, x \neq 0$$

and that $F'(0) = d < 0$.

See [14] for an interesting study of a related system wherein it is shown that it exhibits complex and chaotic behavior for suitable parameter values. We are motivated here by the paper of Cooke and Grossman [20].

In system form, our equation becomes

$$\begin{aligned} x'(t) &= v(t) \\ v'(t) &= -bv(t) - ax(t) + F(x(t - r)) \end{aligned} \tag{6.19}$$

Its only steady-state is $x = 0$ and the corresponding characteristic equation is

$$\lambda^2 + b\lambda + a = de^{-\lambda r} \tag{6.20}$$

It is useful later to scale time so the delay becomes one. The scaling

$$X(\tau) = x(t), V(\tau) = v(t), r\tau = t$$

results in

$$\begin{aligned} \dot{X}(\tau) &= rV(\tau) \\ \dot{V}(\tau) &= -brV(\tau) - arX(\tau) + rF(X(\tau - 1)) \end{aligned}$$

Lemma 6.1. *There are no real roots satisfying $\lambda \geq 0$.*

Proof. The graph of $g(\lambda) := \lambda^2 + b\lambda$, a parabola, cannot meet the graph of $h(\lambda) :=$ $-a + de^{-r\lambda}$ for $\lambda \geq 0$ inasmuch as $h' > 0$, $h(0) = d - a < 0$, and $h(\infty) = -a < 0$. Draw a picture. $\square$

Lemma 6.2. *There exists $R > 0$, depending only on a, b, d, such that if λ is a root with $\Re(\lambda) \geq 0$, then $|\lambda| < R$.*

Proof. Let $f(\lambda, r) = \lambda^2 + b\lambda + a - de^{-\lambda r}$. Then

$$\frac{f(\lambda, r)}{\lambda^2} = 1 + \lambda^{-2}[b\lambda + a - de^{-r\lambda}]$$

Because $|e^{-r\lambda}| \leq 1$ if $\Re(\lambda) \geq 0$, there exists $R > 0$ such that the second summand on the right is less than $1/2$ in modulus when $\Re(\lambda) \geq 0$ and $|\lambda| \geq R$. This proves our assertion. $\square$

Define

$$M(r) = \#\{\lambda : \Re(\lambda) \geq 0 \text{ and } f(\lambda, r) = 0\}$$

where we count according to multiplicity the roots with nonnegative real part.

The following result says that the only way that $M(r) \neq M(r')$ for $r < r'$ is for there to be a purely imaginary root for some r'' between r and r'. It is adapted from [20].

Proposition 6.2 *Let $0 \leq r_1 < r_2$. Suppose that for $r_1 \leq r \leq r_2$ there are no roots of (6.20) on the imaginary axis. Then*

$$M(r_1) = M(r_2)$$

Proof. Let γ be the Jordan curve consisting of the semicircle $|\lambda| = R$ with $\Re(\lambda) \geq 0$ and the portion of the imaginary axis between $\pm iR$, oriented counterclockwise. By Lemma 6.2, any root with $\Re(\lambda) \geq 0$ lies inside γ. Let

$$m = \min\{|f(\lambda, r)| : \lambda \in \gamma, r \in [r_1, r_2]\}$$

By hypothesis and Lemma 6.2, $|f(\lambda, r)| > 0$ for $\lambda \in \gamma$ and $r_1 \leq r \leq r_2$ so $m > 0$ by continuity of $(\lambda, r) \to |f(\lambda, r)|$ and compactness of $\gamma \times [r_1, r_2]$.

By uniform continuity of $(\lambda, r) \to f(\lambda, r)$ on the compact set $\gamma \times [r_1, r_2]$, there exists $\delta > 0$ such that

$$|f(\lambda, r) - f(\lambda, r')| < m/2, r, r' \in [r_1, r_2], \lambda \in \gamma, |r - r'| \leq \delta.$$

By Theorem A.4,

$$M(r) = M(r'), r, r' \in [r_1, r_2], |r - r'| \leq \delta.$$

Therefore $M(r_1) = M(r_2)$ because we can take "steps" of length δ. $\quad\square$

Remark 6.3 *Examination of the proof of Proposition 6.2 indicates that it is applicable to general analytic functions depending on a real parameter, or parameters, where an a priori bound on roots as in Lemma 6.2 can be obtained that is independent of the parameters that are being varied. See [64] for a general result.*

Now consider roots $\lambda = i\omega$ where $\omega > 0$. We must have

$$a - \omega^2 + ib\omega = de^{-i\omega r}$$

so

$$a - \omega^2 = d\cos(\omega r)$$
$$b\omega = -d\sin(\omega r)$$

Squaring both sides and adding gives

$$\omega^4 + (b^2 - 2a)\omega^2 + a^2 - d^2 = 0$$

Define the discriminant by

$$\triangle = (b^2 - 2a)^2 - 4(a^2 - d^2)$$

Solving for ω^2, we have

$$\omega_\pm^2 = \frac{1}{2}[2a - b^2 \pm \sqrt{\triangle}] \tag{6.21}$$

As we must have $\omega > 0$, the following cases emerge.

(1) $\triangle < 0$ implies no purely imaginary roots.
(2) $\triangle > 0$, $2a - b^2 < 0$, $a^2 > d^2$ implies no purely imaginary roots.
(3) $\triangle > 0$, $2a - b^2 < 0$, $a^2 < d^2$ implies one root $i\omega_+$.
(4) $\triangle > 0$, $2a - b^2 > 0$, $a^2 < d^2$ implies one root $i\omega_+$.
(5) $\triangle > 0$, $2a - b^2 > 0$, $a^2 > d^2$ implies two roots $i\omega_\pm$.

6.4.1 Delayed Feedback Dominates Instantaneous Feedback

Having determined ω, we must now determine r. We consider cases (3) and (4) together. Then $a^2 < d^2$ and

$$\omega_+ = \sqrt{\frac{2a - b^2 + \sqrt{\triangle}}{2}}$$

and

$$\cos(\omega r) = (1/2|d|)[-b^2 + \sqrt{\triangle}]$$

$$\sin(\omega r) = -\frac{b}{d}\,\omega_+ > 0 \tag{6.22}$$

Thus

$$r_n = \frac{1}{\omega_+}[\sin^{-1}(-\frac{b}{d}\,\omega_+) + 2n\pi],\ n = 0,1,2,\dots$$

where the branch of $\sin^{-1}$ depends on the sign of term opposite $\cos(\omega r)$ above. If $(1/2|d|)[-b^2 + \sqrt{\triangle}] < 0$ then we take the inverse of the decreasing branch $\sin : (\pi/2, 3\pi/2) \to (-1,1)$; if $(1/2|d|)[-b^2 + \sqrt{\triangle}] > 0$ then we take the inverse of the increasing branch $\sin : (-\pi/2, \pi/2) \to (-1,1)$.

Multiplying $[-b^2 + \sqrt{\triangle}]$ by $[b^2 + \sqrt{\triangle}]$ doesn't change the sign of the former and results in $4(d^2 - a b^2)$. Note $4(d^2 - ab^2) > 4(a^2 - ab^2) = 4a(a - b^2)$.

If $d^2 < ab^2$ then $[-b^2 + \sqrt{\triangle}] < 0$ so $\pi/2 < \sin^{-1}(-b\omega_+/d) < \pi$ so the smallest delay for which there is a purely imaginary pair of roots $\pm i\omega_+$ is

$$r_0 = \frac{1}{\omega_+}\sin^{-1}(-\frac{b}{d}\,\omega_+).$$

If $d^2 > ab^2$ then $[-b^2 + \sqrt{\triangle}] > 0$ so $0 < \sin^{-1}(-b\omega_+/d) < \pi/2$ so the smallest delay for which there is a purely imaginary pair of roots $\pm i\omega_+$ is

$$r_0 = \frac{1}{\omega_+}\sin^{-1}(-\frac{b}{d}\,\omega_+).$$

Proposition 6.4 *Assume that* $\triangle > 0$, $a^2 < d^2$, *and* $0 \le r < r_0$. *Then* $\Re(\lambda) < 0$ *for every root of* (6.20). $x = 0$ *is asymptotically stable.*

Proof. By Proposition 6.2, $M(r) = M(0) = 0$. $\square$

Now let's see if a Hopf bifurcation occurs at $r = r_0$. Differentiating implicitly (6.20) with respect to r we find that

$$2\lambda\frac{d\lambda}{dr} + b\frac{d\lambda}{dr} = -de^{-\lambda r}(r\frac{d\lambda}{dr} + \lambda)$$

Substituting $\lambda = i\omega_+$ we find that

$$\frac{d\lambda}{dr}(r) = \frac{-i\omega_+ d\cos(\omega_+ r) - \omega_+ d\sin(\omega_+ r)}{b + dr\cos(\omega_+ r) + i(2\omega_+ - rd\sin(\omega_+ r))} \tag{6.23}$$

Using (6.22), multiplying top and bottom by the conjugate of the denominator, and taking real part, we get

$$\frac{d\Re(\lambda)}{dr}(r_0) = \frac{\sqrt{\triangle}\,\omega_+^2}{|\text{denominator}|^2} \tag{6.24}$$

where

$$|\text{denominator}|^2 = [b + (r_0/2)(b^2 - \sqrt{\triangle})]^2 + \omega_+^2(2 + r_0 b)^2 > 0$$

The fact that the above is positive (not zero) ensures the hypotheses of the implicit function theorem hold so we may conclude that for $r \approx r_0$ there is a root

$$\lambda = \lambda(r) = \alpha(r) + i\omega(r), \alpha(r_0) = 0,\ \alpha'(r_0) = \frac{d\Re(\lambda)}{dr}(r_0) > 0, \omega(r_0) = \omega_+$$

This root crosses the imaginary axis at $r = r_0$ from left to right in the sense that $\alpha < 0$ for $r < r_0$ and $\alpha > 0$ when $r > r_0$, at least when $r \approx r_0$. Actually, our calculation used no information about r_0; the same conclusion holds at any $r = r_n$ where there is a purely imaginary root $i\omega_+$. Consequently, we conclude the following.

Proposition 6.5 *Assume that* $\triangle > 0$, $a^2 < d^2$. *For* $r > r_0$, $x = 0$ *is unstable.*

Proof. Certainly for $r > r_0$ with $r - r_0$ small, the implicit function theorem (IFT) implies that there is a complex conjugate pair of roots with positive real part. There are no other roots with positive real part. To make this point clear, recall that the IFT implies there is a neighborhood U about $i\omega_+$ and a small open interval I containing r_0 such that the only root of the characteristic equation in U for $r \in I$ is $\alpha(r) + i\omega(r)$, the one given by IFT. We may take U to be a disk of radius $s > 0$. Now let's apply Rouché's theorem using the Jordan curve Γ which is the same as γ except that when the vertical part of γ meets U it traces the right half of the disk bounding U. Then we can see that, by shrinking I if necessary, for $r \in I$, there are no roots on Γ and $|f(\lambda, r) - f(\lambda, r_0)| < |f(\lambda, r_0)|$ for $\lambda \in \Gamma$, $r \in I$. Consequently the sum of the multiplicity of the roots inside Γ cannot change as r varies in I. Because that number is 0 for $r < r_0$, $r \in I$, by the previous proposition, it follows that there are no roots in the right half-plane other than the one inside U for $r \geq r_0$, $r \in I$. Hence, for such r, the number of roots inside γ is $M(r) = 2$; furthermore, by Proposition 6.2, $M(r) = 2$ as long as $r_0 \leq r < r_1$. For $r \approx r_1$, the IFT again implies that a pair of roots crosses the imaginary axis:

$$\lambda(r) = \alpha_*(r) + i\omega_*(r), \alpha_*(r_1) = 0,\ \alpha'_*(r_1) > 0, \omega_*(r_1) = \omega_+$$

Arguing as above using a supplementary Jordan curve Γ that contains two roots inside it this time, and using Proposition 6.2, we can see that $M(r) = 4$ for $r_1 \leq r < r_2$. Induction establishes that $M(r) = 2n$ for $r_{n-1} \leq r < r_n$. Consequently, for all $r > r_0$ there exists at least one root with positive real part. $\square$

We have verified all the hypotheses of the Hopf bifurcation theorem.

Proposition 6.6 *Assume that* $\triangle > 0$, $a^2 < d^2$, *and that* F *is* C^2. *Then a Hopf bifurcation occurs at* $r = r_0$.

6.4.2 Instantaneous Feedback Dominates Delayed Feedback

Now we study the case: $a^2 > d^2$, $2a - b^2 > 0$.

This is case (5) above. In this case there are two purely imaginary roots $i\omega$ with $\omega > 0$, namely:

$$\omega_\pm = \sqrt{\frac{2a - b^2 \pm \sqrt{\triangle}}{2}}$$

satisfying $0 < \omega_- < \omega_+$. To determine for which values of r these are roots, we proceed as before

$$\cos(\omega_\pm r) = (1/2|d|)[-b^2 \pm \sqrt{\triangle}]$$
$$\sin(\omega_\pm r) = -\frac{b}{d}\,\omega_\pm > 0 \tag{6.25}$$

Because $\triangle < (b^2 - 2a)^2 < b^4$ we see that $[-b^2 \pm \sqrt{\triangle}] < 0$. Thus

$$r_n^\pm = \frac{1}{\omega_\pm}[\sin^{-1}(-\frac{b}{d}\,\omega_\pm) + 2n\pi], n = 0, 1, 2, \ldots$$

where we use the decreasing branch $\sin : (\pi/2, 3\pi/2) \to (-1, 1)$ because $\cos(\omega_\pm r) < 0$. For simplicity, define

$$\theta_\pm = \sin^{-1}(-\frac{b}{d}\,\omega_\pm)$$

Note that $\pi/2 < \theta_- < \theta_+ < \pi$ inasmuch as $0 < \omega_- < \omega_+$ and $\sin^{-1}$ is decreasing. So we have two sequences of delays

$$r_n^+ = \frac{\theta_+ + 2n\pi}{\omega_+}, r_n^- = \frac{\theta_- + 2n\pi}{\omega_-}$$

for which there are purely imaginary roots $\pm i\omega_+$, respectively, $\pm i\omega_-$.

The implicit function theorem applies at each of these roots and the formula (6.23) holds for ω_+ as before so we have for $\lambda_+(r) = \alpha_+(r) + i\omega_+(r)$, $\alpha_+(r_n^+) = 0$, $\omega_+(r_+^n) = \omega_+$ and

$$\frac{d\alpha_+}{dr}(r_n^+) > 0$$

as before. However at $r = r_n^-$, where $i\omega_-$ is a root, and where the implicit function theorem also applies so $\lambda_-(r) = \alpha_-(r) + i\omega_-(r)$, $\alpha_-(r_n^-) = 0$, $\omega_-(r_-^n) = \omega_-$, the calculation that gave (6.24) now gives

$$\frac{d\alpha_-}{dr}(r_n^-) = \frac{-\sqrt{\triangle}\,\omega_-^2}{|\text{denominator}|^2} < 0$$

Therefore this family of roots crosses in the opposite direction as r increases through r_n^-.

Consider the special case $a = 3, b = 1, d = -\sqrt{5}$ so that

$$\omega_+ = 2, \omega_- = 1, \theta_+ = 2.03, \theta_- = 2.68,$$

the last two values being approximations. Then

$$r_n^+ = 1.02 + n\pi$$
$$r_n^- = 2.68 + 2n\pi$$

so we have

$$r_0^+ < r_0^- < r_1^+ < r_2^+ < r_1^- < \cdots$$

As r is increased, starting from $r = 0$, the first pair of complex conjugate roots crosses from the left half-plane into the right half-plane at $r = r_0^+$ at $i\omega_+$ but then when $r = r_0^-$ this same pair crosses from the right half-plane back into the left half-plane, at $i\omega_-$. Another, not necessarily the same, pair of roots then crosses into the right half-plane at $i\omega_+$ when r_1^+. Now it becomes hard to follow the action because yet another pair of roots crosses from the left half-plane into the right half-plane at $i\omega_+$ when $r = r_2^+$ after which at r_1^- a pair crosses in the opposite direction at $i\omega_-$. It is now clear that as r is further increased, roots cross from left to right more frequently (as r increases) than roots cross in the opposite direction. Therefore, we have the following conjecture.

Proposition 6.7 *Let $a = 3, b = 1, d = -\sqrt{5}$. Then $x = 0$ is asymptotically stable for $0 \le r < r_0^+$, unstable on $r_0^+ < r < r_0^-$, asymptotically stable for $r_0^- < r < r_1^+$, and unstable for $r > r_1^+$.*

Proof. We argue that $0 = M(0) = M(r)$ for $0 \le r < r_0^+$ as in the previous case. Similarly, $2 = M(r_0^+) = M(r)$ for $r_0^+ \le r < r_0^-$. But now we want to show that this pair of roots crosses back into the left half-plane when $r = r_0^-$. We argue in a similar way as in Proposition 6.5. We have a root $i\omega_-$ at $r = r_0^-$ and the implicit function theorem implies the existence of a closed disk centered at $i\omega_-$ and an open interval I about $r = r_0^-$ such that the only root of the characteristic equation in U for $r \in I$ is the root $\lambda_-(r) = \alpha_-(r) + i\omega_-(r)$, where $\alpha_-(r_0^-) = 0, \omega_-(r_0^-) = \omega_-$ and $\alpha_-(r) < 0$ for $r > r_0^-$, $\alpha_-(r) > 0$ for $r < r_0^-$, $r \in I$. By shrinking I, if necessary, we can assume that $\lambda_-(r)$ does not lie on the boundary of U for $r \in I$. Let Γ be the Jordan curve which is the same as that in Proposition 6.5 except that instead of following the right half of the circular boundary of U, it follows the left half of the circle. By shrinking I again, if necessary, we can assume that the hypotheses of Rouché's theorem hold: $|f(\lambda, r_0^-) - f(\lambda, r)| < |f(\lambda, r_0^-)|$ for $r \in I$ and $\lambda \in \Gamma$. Therefore, the number of roots, counting multiplicity, of $f = 0$ inside Γ is the same for all $r \in I$. For $r < r_0^-$, that number is 2 because $M(r) = 2$ and $\alpha_-(r) > 0$. So it must also be 2 for $r > r_0^-$, $r \in I$. But for such r, $\alpha_-(r) < 0$, so $\lambda_-(r)$ lies in the open left half-plane. It follows that $M(r) = 0$ for $r \in I, r > r_0^-$.

By Proposition 6.2, it follows that $M(r) = 0$ for $r_0^- < r < r_1^+$. The argument that $M(r) = 2$ for $r_1^+ \le r < r_2^+$ is identical to previous arguments, as is that for $M(r) = 4$

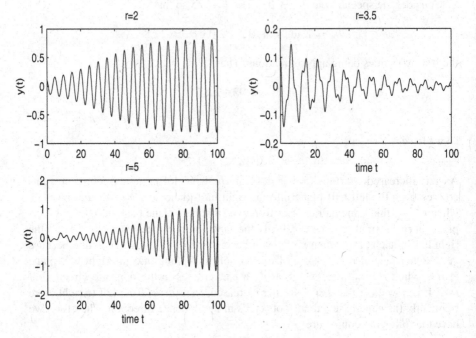

Fig. 6.3 Simulations of $x'' + x' + 3x = -\sqrt{5}\tanh(x(t-r))$ for various r.

for $r_2^+ \le r < r_1^-$, and $M(r) = 2$ for $r_1^- \le r < r_3^+$. The remainder of the proof is more of the same. $\square$

In Figure 6.3 we simulate the second-order equation

$$x'' + x' + 3x = -\sqrt{5}\tanh(x(t-r))$$

using three values of r. The first $r = 2$ is between r_0^+ and r_0^-, the second $r = 3.5$ lies between r_0^- and r_1^+, and the last, $r = 5$ is greater than r_1^+.

6.4.3 Stabilizing the Straight-Up Steady State of the Pendulum

Consider the damped pendulum equation where we have added a delayed negative state feedback restoring force to try to stabilize the straight-up equilibrium $\theta = \pi$:

$$ml\theta''(t) + k\theta'(t) + mg\sin(\theta(t)) = R(\pi - \theta(t-r))$$

θ represents the (counterclockwise) angle from the straight-down position. Parameter m denotes the mass on the end of the pendulum of length l; g is the gravitational constant and k is a measure of the damping. Recall that when $R = 0$, the "down" steady-state $\theta = 0$ is asymptotically stable if $k > 0$ and the "up" steady state $\theta = \pi$ is an unstable saddle point. Can we choose the "gain" $R > 0$ so as to stabilize the up steady-state if there are delays in its implementation? This could be viewed as a cartoon model for maintaining a yardstick straight-up while balanced on the palm of the hand by moving the hand back and forth. For a more realistic model, see [49].

As we are interested in the up steady state, we change variables

$$\theta = y + \pi$$

so that the up steady state $\theta = \pi$ becomes $y = 0$. The equation becomes

$$mly''(t) + ky'(t) - mg\sin(y(t)) = -Ry(t - r)$$

Setting

$$v^2 := g/l, \ d := k/ml, \ p := R/ml$$

we get

$$y''(t) + dy'(t) - v^2\sin(y(t)) = -py(t - r) \tag{6.26}$$

If $k = 0$ the ordinary pendulum oscillates about the down steady state with frequency v and period $2\pi/v = 2\pi\sqrt{l/g}$ so the natural time scale of the undamped pendulum depends on the length of the pendulum l. The other time scale in our problem is the delay r in implementing the restoring force.

Linearizing about the $y = 0$ steady state, we obtain

$$Y''(t) + dY'(t) - v^2Y(t) = -pY(t - r)$$

which leads to the characteristic equation

$$f(\lambda) = f(\lambda, r, p) := \lambda^2 + d\lambda - v^2 + pe^{-\lambda r} = 0$$

Our goal is to find the stability region in the (r, p)-plane and to determine the behavior of solutions near the steady state $(y, y') = (0, 0)$ as parameters (r, p) are varied near the boundary of the stable region. As usual, we expect that the stability region is bounded by curves along which 0 is a root and curves along which $\pm i\omega$ is a root for some ω.

Before plunging on, it is a good idea to consider the case $r = 0$ where there is no delay in implementing the restoring force. In that case, the characteristic equation is the quadratic $\lambda^2 + d\lambda + p - v^2 = 0$. Both roots have negative real part when $p - v^2 > 0$; if $p - v^2 < 0$ there is a positive and a negative root so $y = 0$ is unstable. In the absence of an implementation delay, the straight-up position of the pendulum can be stabilized if p exceeds v^2, that is, if R is sufficiently large.

Let's first consider real roots. Observe that $f(\pm\infty) = +\infty$ and $f(0) = p - v^2$ so 0 is a root when $p = v^2$. As $f'(0) = d - pr$, it is an order-one root when $r \neq d/v^2$. Because $f''(0) = 2 - pr^2$, 0 is a root of multiplicity two when $p = v^2$, $r = d/v^2$

and $d^2/v^2 \neq 2$. Obviously, if $p - v^2 < 0$ then there are at least two real roots, one of which is positive. Thus $p < v^2$ is a region of instability.

For $\lambda = i\omega$, $\omega > 0$ to be a root we must have

$$-\omega^2 + di\omega - v^2 + pe^{-i\omega r} = 0$$

or

$$\omega^2 + v^2 = p\cos(\omega r)$$
$$d\omega = p\sin(\omega r)$$

Squaring both sides determines the value of p as a function of ω:

$$(\omega^2 + v^2)^2 + d^2\omega^2 = p^2$$

The smallest value of the delay r is determined in the usual way by the equation for $\sin(\omega r)$ above, so we find that $i\omega$ is a root along the curve

$$r = \frac{1}{\omega}[\sin^{-1}(d\omega/\sqrt{(\omega^2 + v^2)^2 + d^2\omega^2})]$$
$$p = \sqrt{(\omega^2 + v^2)^2 + d^2\omega^2}, \ \omega > 0 \qquad (6.27)$$

where the inverse sine is determined by $\sin : (-\pi/2, \pi/2) \to (-1, 1)$ because both sine and cosine are positive. This curve is plotted in Figure 6.4 where $d = v^2 = 1$. As expected, when $\omega \searrow 0$, the point $(r, p) \to (d/v^2, v^2)$, the point at which 0 is a double root. The curve begins there and goes to infinity tangent to the p-axis following the hyperbola $pr = d$. This curve along which $i\omega$ is a root and the curve $p = v^2$ along which 0 is a root form the boundary of the nose-shaped stability region.

Proposition 6.8 *The region of asymptotic stability for the zero solution of (6.26) is the "nose"-shaped region bounded by the lines $r = 0$, $p = v^2$ and by the curve (6.27).*

A Hopf bifurcation occurs as r is increased such that (r, p) crosses the curve (6.27); A steady-state bifurcation occurs as p is decreased such that (r, p) crosses the line $p = v^2$.

Proof. Note that when $r = 0$ and $p > v^2$, both roots of the quadratic have negative real part. Also, as noted above there is a positive root when $p < v^2$ below the nose. Along the curve (6.27) there is a purely imaginary root $i\omega$. A standard implicit function theorem argument shows that

$$\Re(\frac{d\lambda}{dr}) = \Re(-\frac{f_r}{f_\lambda}) = \frac{d^2\omega^2 + 2\omega^2(\omega^2 + v^2)}{|f_\lambda|^2} > 0$$

implying that the $\Re(\lambda(r)) > 0$ for $r > r(\omega)$ at least for r near $r(\omega)$. Therefore, just below and to the right of the nose, there are roots with positive real part. The

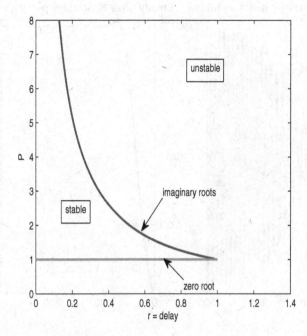

Fig. 6.4 Stability region lies inside the "nose", here, drawn for case: $d = v^2 = 1$.

hypotheses of Theorem 6.1 are satisfied so we conclude that a Hopf bifurcation occurs as we increase r such that (r, p) moves across the curve (6.27).

Steady-states $(y, y') = (y, 0)$ for (6.26) must satisfy

$$g(y, p) \equiv py - v^2 \sin(y) = 0 \tag{6.28}$$

Clearly, $g(0, p) = 0$ and $(\partial g / \partial y)(0, p) = p - v^2$. By the implicit function theorem, there can be no solutions of (6.28) sufficiently near $(0, p)$ other than $(0, p)$ if $p \neq v^2$. However, we might expect a bifurcation from the "branch of trivial steady states" $\{(y, p) = (0, p) : p \in \mathbb{R}\}$ to occur near $(0, v^2)$ because the implicit function theorem fails at this point.

Now we apply Rouché's theorem, exactly as we did in Proposition 4.6 for the delayed logistic equation, in order to show that the number of roots satisfying $\Re(\lambda) \geq 0$ is a continuous function of (r, p) inside this region and on the $r = 0$ part of it. Therefore, this number must be zero in the region because it is 0 on the $r = 0$ part of it. A key to this argument is that no roots appear in the imaginary axis except along the line $p = v^2$, on the curve (6.27), and along the curves obtained from (6.27) by adding $n\pi/\omega$ to the r part. These curves lie to the right of our region; notice that as $\omega \searrow 0$ along these curves that $r(\omega) \nearrow \infty$. $\square$

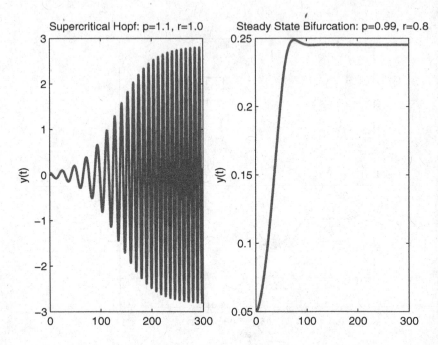

Fig. 6.5 Simulations of $y'' + y' - \sin(y) = -py(t - r)$.

In summary, p must exceed v^2 in order to stabilize the up steady state and the delay r cannot exceed one. Indeed, if $0 < r < d/v^2$ is fixed and p is increased starting from $p = v^2$, the point (r, p) remains in the stability region for an interval of p values before leaving the stability region through the "bridge of the nose."

Note that the corner of the nose occurs at $(r, p) = (d/v^2, v^2)$, where 0 is a double root. Returning to the unscaled parameters $p = R/ml$, $d = k/ml$, $v^2 = g/l$ we see that the requirement that $p > v^2$ for stabilization becomes $R > mg$. The delay r must necessarily satisfy $r < d/v^2 = k/mg$ (but this is not sufficient!) in order to stabilize the up steady state; this threshold may be very small for small damping!

Figure 6.5 provides two simulations for values of (r, p) just outside the stability region. The first $(r, p) = (1, 1.1)$ is just above and to the right of the tip of the nose, which occurs at $(1, 1)$; it clearly shows stable oscillations reflecting a supercritical Hopf bifurcation. The second, at $(r, p) = (0.8, 0.99)$, just below the nose, shows the existence of a stable nonzero steady state. The initial data for both cases are constant: $y = 0.05$, $y' = 0$.

6.5 Gene Regulation by End-Product Repression

Consider the negative feedback gene regulatory system

$$x_1'(t) = g(x_2(t-r)) - \alpha_1 x_1(t)$$
$$x_2'(t) = x_1(t) - \alpha_2 x_2(t) \tag{6.29}$$

where x_1 denotes intracellular mRNA and x_2 denotes the protein product of the gene. The delay r represents the time for mRNA to leave the nucleus where it is translated into protein in the ribosome, whereupon the protein re-enters the nucleus and represses the production of its own mRNA: $g' < 0$. We assume that $g : [0, \infty) \to [0, \infty)$ is C^1. An example is given by the Hill function

$$g(x) = g_m \frac{1}{1 + (x/K)^p}$$

For more background on gene regulatory modeling, see [32, 3, 68].

We expect (6.29) to generate a semiflow on $C_+ = \mathbb{R}_+ \times C([-r, 0], \mathbb{R}_+)$. Theorem 2.4 implies that solutions corresponding to nonnegative initial data remain nonnegative. Boundedness of solutions starting in C_+ will establish that solutions of initial-value problems extend to $[0, \infty)$. This follows from differential inequality arguments as any solution satisfies

$$x_1' \leq g(0) - \alpha_1 x_1$$

and hence $x_1(t)$ is bounded above by the solution $v(t)$ of the linear differential equality satisfying $v(0) = x_1(0)$. Because x_1 is bounded, x_2 must be as well. Hence solutions extend to the infinite interval.

The equilibrium solution is given by

$$E = (\alpha_2 u, u)$$

where $u > 0$ is the unique solution of $g(u) = \alpha_1 \alpha_2 u$. Linearizing about it, we get the linear system (4.29) where matrices A and B are given by

$$A = \begin{pmatrix} -\alpha_1 & 0 \\ 1 & -\alpha_2 \end{pmatrix}, B = \begin{pmatrix} 0 & g'(u) \\ 0 & 0 \end{pmatrix}$$

The characteristic equation is

$$(\alpha_1 + \lambda)(\alpha_2 + \lambda) - g'(u)e^{-r\lambda} = 0 \tag{6.30}$$

Note that (6.30) agrees exactly with (6.20) studied earlier.

If $r = 0$ the resulting quadratic has positive coefficients and so both roots have negative real part. Hence, for small values of the delay, the equilibrium is asymptotically stable by Theorem 6.8. Note also that the divergence of the vector field is $-(\alpha_1 + \alpha_2) < 0$ when the delay is absent. Therefore, by the Dulac theorem [78], no periodic solutions exist if $r = 0$ and equilibrium E is globally asymptotically stable.

Corollary 4.10 can also be applied to (6.30) where $c = g'(u)$. If

$$\frac{g'(u)}{\alpha_1 \alpha_2} > -1 \qquad (6.31)$$

then the equilibrium is asymptotically stable for all values of the delay.

Let $G : [0, \infty) \to [0, \infty)$ be defined by

$$G(x) = \frac{g(x)}{\alpha_1 \alpha_2}$$

There is a remarkable connection between the discrete dynamics of the map G and that of system (6.29), noted by Allwright [3] and extended in [68]. Observe that $G(u) = u$ and that (6.31) implies that u is an asymptotically stable fixed point. Because $G^2 = G \circ G$ is monotone increasing, it follows that every orbit $\{u_n = G^n(u_0)\}$ is asymptotic to a period-two orbit. If the fixed point u is the only period-two orbit, then it is globally attracting for G. The following result establishes that global stability of the fixed point u for G implies global stability of E. See [68] for extensions of the result.

Theorem 6.9 *Suppose the map G has no period-two point except the fixed point u. Then equilibrium E is globally asymptotically stable for (6.29).*

Proof. Let $x_i^\infty = \limsup_{t \to \infty} x_i(t)$ and $x_{i\infty} = \liminf_{t \to \infty} x_1(t)$. From

$$x_1' \leq g(0) - \alpha_1 x_1$$

we deduce that $x_1^\infty \leq g(0)/\alpha_1$. Using this in the equation for x_2 implies that for $\varepsilon > 0$

$$x_2' \leq (x_1^\infty + \varepsilon) - \alpha_2 x_2$$

holds for all large t depending on $\varepsilon > 0$. Thus $x_2^\infty \leq g(0)/(\alpha_1 \alpha_2) = G(0)$.

Now we turn these differential inequalities around. As

$$x_1' \geq g(G(0) + \varepsilon) - \alpha_1 x_1$$

for large t, depending on ε, it follows that $x_{1\infty} \geq g(G(0) + \varepsilon)/\alpha_1$ and, inasmuch as $\varepsilon > 0$ was arbitrary and small, $x_{1\infty} \geq g(G(0))/\alpha_1$. Then

$$x_2' \geq g(G(0))/\alpha_1 - \varepsilon - \alpha_2 x_2$$

for all large t, depending on ε so $x_{2\infty} \geq G^2(0)$. Summarizing the results so far, we have

$$G^2(0) \leq x_{2\infty} \leq x_2^\infty \leq G(0)$$

We now iterate this argument. Starting with $x_2(t - r) \geq G^2(0) - \varepsilon$, which holds for large t, we have

$$x_1' \leq g(G^2(0) - \varepsilon) - \alpha_1 x_1$$

We conclude that $x_1^\infty \leq g(G^2(0))/\alpha_1$ and then, using the second equation, that $x_2^\infty \leq G^3(0)$. Now reversing the inequalities, as above, leads to:

$$G^2(0) \leq G^4(0) \leq x_{2\infty} \leq x_2^\infty \leq G^3(0) \leq G(0)$$

Here, we have made use of cobwebbing to obtain the relative order of the iterates $\{G^n(0)\}$.

An induction argument yields

$$G^{2n}(0) \leq x_{2\infty} \leq x_2^\infty \leq G^{2n-1}(0)$$

As $G^n(0) \to u$, we conclude that $\lim x_2(t) = u$ and from this we easily conclude that $\lim x_1(t) = \alpha_2 u$. $\square$

Sufficient conditions for G to have no period-two points other than u are given in Corollary 9.9 of [75]. One of these is that $xG(x)$ is strictly monotone on $(0, G(0))$. See Remark 9.10 [75]. An easy computation shows that, for the Hill function, $xG(x)$ is strictly increasing on this interval provided

$$(p-1)\left[\frac{gm}{K\alpha_1\alpha_2}\right]^p < 1$$

If (6.31) fails, that is, if

$$G'(u) < -1 \qquad (6.32)$$

then the fixed point u is unstable and G has period-two points distinct from u so the Theorem 6.9 no longer applies. Theorem 6.9 is only useful when E is asymptotically stable for all delays. Results below and our simulations using the Hill function show that convergence to equilibrium can occur when the hypotheses of Theorem 6.9 fail.

Hereafter, we restrict attention to the case that (6.32) holds and make use of results concerning the characteristic equation (6.20). Using the notation of Section 6.4 ($b = \alpha_1 + \alpha_2, a = \alpha_1\alpha_2, d = g'(u)$), we have

$$\triangle = (\alpha_1^2 - \alpha_2^2)^2 + 4(g'(u))^2 > 0$$
$$2a - b^2 = -(\alpha_1^2 - \alpha_2^2) < 0$$
$$a^2 - d^2 = (\alpha_1\alpha_2)^2[1 - G'(u)^2] < 0$$

Thus, we are in the case that there exists a single purely imaginary root $\lambda = \pm i\omega_+$ given by (6.21):

$$\omega_+ = \sqrt{\frac{2a - b^2 + \sqrt{\triangle}}{2}}$$

This root corresponds to an increasing sequence of values of the delay r. The smallest such $r = r_0$ is identified below. According to Proposition 6.4 and Proposition 6.5, we have the following result.

Proposition 6.10 *Assume that (6.32) holds. If $0 \leq r < r_0$, where*

$$r_0 = \frac{1}{\omega_+} \sin^{-1}\left(-\frac{b}{d}\,\omega_+\right)$$

then E is locally asymptotically stable. For $r = r_0$, there is a simple pair of purely imaginary roots $\pm i\omega_+$. If $r > r_0$, then there exists at least one pair of complex characteristic roots with positive real part. Hence, E is unstable. A Hopf bifurcation occurs as r passes through r_0.

Convergence of solutions to the equilibrium E may hold even if the map G has a period-two orbit as Figure 6.6 and Figure 6.8 show. However, if the delay is too large, sustained oscillations may occur as shown in Figure 6.7.

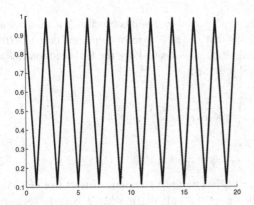

Fig. 6.6 Iterates of the map G with parameters: $K = 0.5$, $\alpha_1 = \alpha_2 = 1$, $p = 3$, $g_m = 1$.

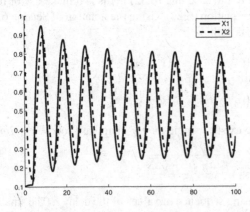

Fig. 6.7 Time series when $r = 3.5$; parameters as above.

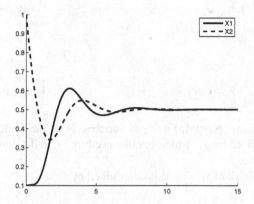

Fig. 6.8 Time series when $r = 0.5$; parameters as above.

6.6 A Poincaré-Bendixson Theorem for Delay Equations

Delay differential equations, even scalar ones, give rise to an infinite-dimensional dynamical system so one should expect extremely complicated dynamics in general. See, for example, [36]. However, special classes of systems can be shown to have simpler dynamics. For example, a certain class of delay equations gives rise to so-called monotone dynamical systems [70, 71] (see Chapter 5) for which it is known that the generic solution converges to equilibrium. A class of systems called monotone cyclic feedback systems (MCFS), which we briefly describe here, enjoy a Poincaré-Bendixson theorem. This was shown for ODEs by J. Mallet-Paret and the author [57] and for delay equations by Mallet-Paret and G. Sell in [56] and a companion paper. The class of delay equations for which the result applies includes the gene regulatory system considered in the previous section as well as systems (1.15), (1.16), and (1.14), to name just a few. In order to simplify the presentation, we restrict attention to a special subclass of MCFS systems treated in [56].

By a MCFS, we mean either the scalar system

$$x'(t) = f(x(t), x(t - r)) \tag{6.33}$$

or the system

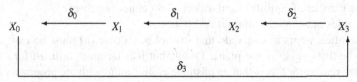

Fig. 6.9 Monotone cyclic feedback cycle: $N = 3$.

$$x_i'(t) = f_i(x_i(t), x_{i+1}(t - r_i)), 0 \leq i \leq N - 1 \tag{6.34}$$
$$x_N'(t) = f_N(x_N(t), x_0(t - r_N))$$

where $r_i \geq 0$, $f, f_i \in C^1$, and

$$\frac{\partial f_i}{\partial y}(x,y) \neq 0, \forall (x,y) \in \mathbb{R}^2 \tag{6.35}$$

Monotonicity is required only in the (possibly) delayed variable. Note the cyclic nature of the system: x_0 is influenced by x_1, which is influenced by x_2, and so on; finally x_N is influenced by $x_{N+1} \equiv x_0$.

The nature of the delayed feedback of x_{i+1} on x_i is determined by

$$\delta_i = \left\{ \begin{array}{l} +1, \ \frac{\partial f_i}{\partial y} > 0 \\ -1, \ \frac{\partial f_i}{\partial y} < 0 \end{array} \right\} \tag{6.36}$$

and the feedback nature of the MCFS is determined by

$$\delta^* = \prod_i \delta_i \tag{6.37}$$

MCFS is said to have positive or negative feedback if $\delta^* = +1$ or $\delta^* = -1$, respectively. System (6.29) has negative feedback. In the case where $\delta^* = +1$, MCFS generates a monotone semiflow and therefore one expects that the generic orbit converges to equilibrium and that attracting periodic orbits cannot exist [44, 70, 71].

Denote by E the set of equilibria of MCFS.

Theorem 6.11 (Mallet-Paret and Sell) *Let $x(t)$ be a bounded solution of the MCFS defined on $[t_0, \infty)$. If $\omega(x)$ denotes its omega limit set, then either*

(a) $\omega(x)$ is a single nonconstant periodic orbit, or
(b) The α and ω limit set of each solution $u(t)$, defined for all $t \in \mathbb{R}$ with $u_t \in \omega(x)$, $t \in \mathbb{R}$, is contained in E.

Case (b) covers the case that $\omega(x)$ is a single equilibrium point. If E consists of isolated points, then the stronger conclusions in (b) that $u(t)$ converges to a single equilibrium as $t \to \infty$ and as $t \to -\infty$ hold; these equilibria may be the same (u is a homoclinic orbit) or different (u is a heteroclinic orbit). By Remark 5.2, through each point $u_0 \in \omega(x)$, there exists a solution $u(t)$ defined for all $t \in \mathbb{R}$ with $u_t \in \omega(x)$, $t \in \mathbb{R}$. Hence, if $E \cap \omega(x)$ consists of isolated points in case (b), then $\omega(x)$ consists of a finite number of equilibria and trajectories connecting them.

Theorem 6.11 rules out chaotic solutions of MCFS.

Of course, one often wants to conclude that (a) holds, so case (b) must be excluded. This can be difficult, even for planar ODEs, but it is far more difficult for delay systems. One reason for this is that an unstable equilibrium, with the property that there are no purely imaginary characteristic roots of the associated characteristic equation, is necessarily a saddle point with an infinite-dimensional stable manifold and a nontrivial finite dimensional unstable manifold. Thus, even if there is only a single unstable equilibrium, such as in the previous section, it is not easy to rule out homoclinic orbits. See [56, 55] for cases where one can rule out homoclinic and heteroclinic orbits, primarily for (6.33).

Many other systems can be converted to the form of MCFS. For example, if we put $y_i = x_{N-i}$, $0 \le i \le N$, then MCFS becomes, on renaming functions and delays, the "feedforward" system

$$y_i'(t) = f_i(y_i(t), y_{i-1}(t - r_i)), 1 \le i \le N \tag{6.38}$$
$$y_0'(t) = f_0(y_0(t), y_N(t - r_N))$$

Mallet-Paret and Sell put (MCFS) in standard form by setting $y_0(t) = x_0(t)$, $y_i(t) = x_i(t - \sum_{j=0}^{i-1} r_j)$, $1 \le i \le N$, which results in a MCFS with single delay $r = \sum_j r_j$:

$$y_i'(t) = f_i(y_i(t), y_{i+1}(t)), 0 \le i \le N-1 \tag{6.39}$$
$$y_N'(t) = f_N(y_N(t), y_0(t - r))$$

Exercises

Exercise 6.1. Determine the stability of the nontrivial branch of steady states of (6.1). Suppose the last two "+" signs in (6.1) are changed to "-" signs. Determine the nontrivial steady-states and their stability in this case.

Exercise 6.2. For $k \ge 0$, let

$$H^k = \{h \in L^2(\mathbb{T}) : \sum_{n \in Z} n^{2k} |h_n|^2 < \infty\}$$

These spaces are Hilbert spaces contained in the Hilbert space $H^0 = L^2(\mathbb{T})$ of square integrable functions on the unit circle $\mathbb{T}$. The larger is k, the smoother are the functions in H^k. Show that if $h \in H^k$ then $P \in H^{k+1}$.

Exercise 6.3. Discuss the direction of bifurcation and stability of the Hopf bifurcation for (6.6) in the case where f is odd.

Exercise 6.4. Consider Hopf bifurcation from the positive equilibrium for the Nicholson blowfly equation (1.3).

Exercise 6.5. Consider the linear second-order equation

$$x''(t) + bx'(t - r) + cx(t) = 0$$

where $b, c > 0$ and $r \ge 0$.

(a) Scale time so the delay becomes one and r is a parameter.
(b) Find the characteristic equation.

(c) Are there real roots satisfying $\lambda \geq 0$?

(d) Find all (r, ω) such that $\pm i\omega$ are roots.

(e) Determine the smallest $r_0 > 0$ for which there is a purely imaginary root.

(f) Verify the hypotheses of the Hopf bifurcation theorem with r_0 as parameter.

Exercise 6.6. Show that there are a pair of nonzero solutions $y = \pm y(s)$ of (6.28) for $p = v^2(1-s)$ for small positive s and that $y(s) \approx \sqrt{6s}$.

Exercise 6.7. If there is no damping ($k = 0$), then there is no stability region for (6.26). Show that if $p > v^2$ and $k = 0$, then there is a pair of purely imaginary roots when $r = 0$. What happens to the real parts of these roots for small $r > 0$?

Exercise 6.8. Show that the Poincaré-Bendixson theorem applies to (1.15) and (1.16).

Exercise 6.9. Show that the Poincaré-Bendixson theorem applies to (6.19) if F is monotone and the roots of the quadratic are real. Hint: Rewrite it as

$$(D - \lambda_1)(D - \lambda_2)x(t) = F(x(t - r))$$

Let $x_0(t) = x(t)$ and $x_1(t) = x'_0(t) - \lambda_1 x_0(t)$.

Exercise 6.10. Make the change of variables $y_0(t) = x_0(t)$, $y_i(t) = (\prod_{j=0}^{i-1} \delta_j) x_i(t - \sum_{j=0}^{i-1} r_j)$, $1 \leq i \leq N$ in MCFS and set $r = \sum_j r_j$. This gives rise to the system:

$$y'_i(t) = g_i(y_i(t), y_{i+1}(t)), 0 \leq i \leq N - 1 \qquad (6.40)$$
$$y'_N(t) = g_N(y_N(t), y_0(t - r))$$

where for $0 \leq i < N$

$$\frac{\partial g_i}{\partial y}(x, y) > 0, \forall (x, y) \in \mathbb{R}^2$$

and

$$\delta^* \frac{\partial g_N}{\partial y}(x, y) > 0, \forall (x, y) \in \mathbb{R}^2$$

Chapter 7
Distributed Delay Equations and the Linear Chain Trick

Abstract Delay differential equations with a special class of distributed infinite delays of gamma type are very attractive both from the modeling point of view and from the point of view of mathematical tractability. A distributed delay is arguably more likely to capture reality than a discrete one. The special class of systems treated here has the feature that by introducing additional components it can be reduced to a system of ODEs. In particular, linearizing about steady states results in a characteristic polynomial, rather than a transcendental equation. We bypass questions of existence and uniqueness and take for granted the validity of linearized stability for these systems. Instead we compare the stability of a distributed delay and a discrete delay in the simplest linear delayed negative feedback equation, showing that distributed delays enhance stability. An example of an HIV transmission model is treated.

7.1 Infinite Delays of Gamma Type

Equations with unbounded distributed delay arise naturally in applications as we saw by examples in the Introduction. For a theoretical treatment the reader may consult Chapter 12 of [41], the book by Hino, Murakami, and Naito [45], and the monograph of Cushing [24]. Here, we introduce some simple ideas, following Mac-Donald [53], which show that a special class of infinite delays leads to ODEs. These delays were already described in the HIV model (1.19) of Culshaw, Ruan, and Webb [23] in Chapter 1.

Basically, we are going to replace the discrete delay

$$y(t-r) = \int_0^\infty y(t-s)\delta(s-r)ds$$

where δ is the Dirac function of unit mass concentrated at zero, in our equations with the distributed delay

H. Smith, *An Introduction to Delay Differential Equations with Applications to the Life Sciences*, 119
Texts in Applied Mathematics 57, DOI 10.1007/978-1-4419-7646-8_7,
© Springer Science+Business Media, LLC 2011

$$\int_0^\infty y(t-s)g_a^p(s)ds = \int_{-\infty}^t y(\eta)g_a^p(t-\eta)d\eta \qquad (7.1)$$

where kernel g_a^p is the density function for a gamma distribution

$$g_a^p(u) := \frac{a^p u^{p-1} e^{-au}}{(p-1)!}, u \geq 0, \qquad (7.2)$$

with parameters $a > 0$ and $p = 1, 2, \ldots$. A nonnegative random variable X is said to obey a gamma distribution if $P(X < c) = \int_0^c g_a^p(u)du$. In particular,

$$\int_0^\infty g_a^p(u)du = 1 \qquad (7.3)$$

Easy calculations show that the mean $\mu = r$ and the variance σ^2 of the gamma distribution are given by

$$r = p/a, \sigma^2 = p/a^2 \qquad (7.4)$$

We view $r = p/a$ as the average delay and σ^2 gives a measure of the degree of concentration of the delay about the mean. Actually, a better measure of the spread of the distribution about the mean for our purposes is the coefficient of variation, the ratio of the standard deviation to the mean: σ/r. The relations (7.4) can be inverted to give:

$$a = r/\sigma^2, p = (r/\sigma)^2 \qquad (7.5)$$

Observe that for the gamma distribution, the reciprocal of the coefficient of variation is restricted to be the square root of a positive integer.

Because g_a^p is a probability distribution, we might view the integral (7.1) as some sort of average value of $y(s)$, $s \leq t$. In particular, the integral lies between $\inf_{s \leq t} y(s)$ and $\sup_{s \leq t} y(s)$.

The discrete delay can be recovered as a limit of gamma-type delays if we let $p = n$, $a = n/r$, $n = 1, 2, \ldots$, so the mean delay remains r but that variance $\sigma^2 = r^2/n \to 0$ as $n \to \infty$. Then we find that

$$\int_0^\infty y(t-s)g_n^{rn}(s)ds \to y(t-r), n \to \infty$$

for any bounded continuous function y. (This can be shown to hold by establishing that the characteristic function of the gamma distribution converges to the characteristic function of the Dirac distribution. See a good book on probability.)

7.1.1 Characteristic Equation and Stability

Consider the simple linear equation

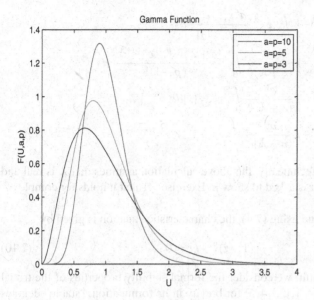

Fig. 7.1 $g_a^p(u)$ for $a = p = 3, 5, 10$. Mean delay is one.

$$y'(t) = -k \int_0^\infty y(t-s) g_a^p(s) ds \qquad (7.6)$$

which we compare with its discrete counterpart considered in Chapter 2:

$$y'(t) = -ky(t-r) \qquad (7.7)$$

We should expect to provide initial data for (7.6) on all of $(-\infty, 0]$

$$y(s) = \phi(s), s \in (-\infty, 0] \qquad (7.8)$$

where ϕ is a bounded and continuous function. Other restrictions on ϕ may be used; see [41, 45]. The initial-value problem (7.6) and (7.8) can be rewritten as

$$y'(t) = \int_0^t y(t-s) k(s) ds + g(t) = (y*k)(t) + g(t) \qquad (7.9)$$

where

$$g(t) = \int_t^\infty \phi(t-s) k(s) ds$$

is a known function and where $y*k$ denotes the convolution of y and k introduced in Section 4.2. Here, we assume that k is integrable and ϕ is bounded and continuous. The Laplace transform is useful for solving such equations. See Exercise 7.2.

A formal stability analysis begins by seeking exponential solutions: $z(t) = e^{\lambda t}$ is a solution of (7.6) if

$$\lambda = -k \int_0^\infty e^{-\lambda s} \frac{a^p s^{p-1} e^{-as}}{(p-1)!} ds$$

$$= -k \frac{a^p}{(a+\lambda)^p} \int_0^\infty \frac{(a+\lambda)^p s^{p-1} e^{-(a+\lambda)s}}{(p-1)!} ds$$

$$= -k \frac{a^p}{(a+\lambda)^p} \int_0^\infty g_{a+\lambda}^p(s) ds$$

$$= -k \frac{a^p)}{(a+\lambda)^p}$$

where we used (7.3). Technically, the above calculation assumes that λ is real and exceeds $-a$. The reader is asked to show in Exercise 7.1 that it holds for complex λ satisfying $\Re\lambda > -a$.

On setting $\lambda = az$ and using (7.4), the characteristic equation is given by

$$pz(1+z)^p + kr = 0 \tag{7.10}$$

In the following result, we consider the formal stability properties of the trivial solution of (7.6) for $p = 1, 2, 3, 4, 5$; for brevity in its formulation, "stable" denotes asymptotic stability and means all roots of (7.10) have negative real part whereas "unstable" means there are roots with positive real part. These formal results can be made rigorous. See Chapter 2 in [24].

Proposition 7.1 *Let $k > 0$. Then there are no nonnegative roots of (7.10). If $p \in \{1,2,3,4,5\}$, the restrictions on kr for stability are summarized in the table:*

p	$1/p = (\sigma/r)^2$	*Stability Condition*
1	*1*	$kr < \infty$
2	*1/2*	$kr < 4$
3	*1/3*	$kr < 8/3 \approx 2.6666$
4	*1/4*	$kr < 224 - 160\sqrt{2} \approx 2.2742$
5	*1/5*	$kr < \frac{16}{5}(7\sqrt{5} - 15) \approx 2.0879$

For each such p, $y = 0$ is unstable if the inequality in the table is reversed.

If $k < 0$, then there is a positive root of (7.10) and hence $y = 0$ is unstable.

Proof. The assertions concerning real roots are left to Exercise 7.3.

First note that when $kr = 0$, roots of (7.10) are $z = 0, -1$ and that as kr becomes positive, the zero root becomes negative (a simple implicit function theorem calculation, see below). Thus, for small $kr > 0$, there are no roots with $\Re z \geq 0$. In addition, roots having nonnegative real part are a priori bounded. Taking the modulus of both sides of $pz(1+z)^p = -kr$ and using $|1+z| \geq |z|$ for $\Re z \geq 0$, we find that $|z|^{p+1} \leq |k|r$. It follows that any roots with nonnegative real part must first appear on the imaginary axis.

In the case where $p = 1$, (7.10) becomes $z^2 + z + kr = 0$ which has no roots with nonnegative real part.

In the other cases $p = 2, 3, 4, 5$, we follow the approach in Proposition 6.2 that justifies considering only purely imaginary roots $z = \pm i\omega$ where $\omega > 0$. For $p = 2$, (7.10) becomes $2z^3 + 4z^2 + 2z + kr = 0$. Plugging in $z = i\omega$, we find that $kr = 4\omega^2$ and $2\omega(1 - \omega^2) = 0$, so $kr = 4$.

It is easily verified that for $p = 3$, (7.10) becomes $3z^4 + 9z^3 + 9z^2 + 3z + kr = 0$ and the substitution yields $3\omega(1 - 3\omega^2) = 0$ and $kr = 3\omega^2(3 - \omega^2) = 8/9$.

Similarly, for $p = 4$, (7.10) becomes $4z^5 + 16z^4 + 24z^3 + 16z^2 + 4z + kr = 0$ and the substitution yields $4\omega(1 - 6\omega^2 + \omega^4) = 0$ and $kr = 16\omega^2(1 - \omega^2)$. Of the two roots, $\omega^2 = 3 - 2\sqrt{2}$ is positive and smaller than one.

If $p = 5$, (7.10) becomes $5z^6 + 25z^5 + 50z^4 + 50z^3 + 25z^2 + 5z + kr = 0$. We find

$$-5\omega^6 + 50\omega^4 - 25\omega^2 + kr = 0$$
$$25\omega^5 - 50\omega^3 + 5\omega = 0$$

Solving the second equation, we find $\omega^2 = 1 - (0.4)\sqrt{5}$ gives $kr > 0$ as above.

In all cases, implicit differentiation of (7.10) yields

$$rk \frac{d(\Re z)}{d(rk)}\Big|_{z=i\omega} = \frac{p\omega^2}{1 + (1 + p)\omega^2} > 0$$

implying that increasing rk from the critical value causes the root $i\omega$ to enter the first quadrant. $\square$

The interesting case is that of delayed negative feedback: $k > 0$. Proposition 7.1 says that as the coefficient of variation decreases, the region of asymptotic stability shrinks. We have only described the first five cases in the countable set of possibilities $(\sigma/r)^2 = 1/n$, $n = 1, 2, 3, \ldots$. These cases should be compared with the stability region

$$0 < kr < \frac{\pi}{2}$$

for the "limiting case" that $(\sigma/r)^2 = 0$ (i.e., the discrete-delay equation (7.7)). Apparently, the trivial solution of the gamma-distributed delay equation (7.6) has a larger region of stability than its discrete counterpart (7.7). This is a general folklore that, quoting from Campbell and Jessop [15]: "A system with a distribution of delays is inherently more stable than the system with a discrete delay."

Remark 7.2 *The proof of Proposition 7.1 shows that purely imaginary roots exist at the right endpoint of the stability window for parameter rk, which implies the existence of a family of periodic solutions of (7.6).*

7.1.2 The Linear Chain Trick

An important property of the functions $g_a^j(u)$, $j = 1, 2, \ldots, p$ is that they satisfy the initial-value problem:

$$\frac{d}{du}g_a^1(u) = -ag_a^1(u), g_a^1(0) = a \tag{7.11}$$

$$\frac{d}{du}g_a^j(u) = a[g_a^{j-1}(u) - g_a^j(u)], g_a^j(0) = 0, \ j = 2, \dots, p$$

Now suppose that $y : (-\infty, T) \to \mathbb{R}$, where $-\infty < T \leq \infty$, is a bounded continuous function and define functions $y_j : (-\infty, T) \to \mathbb{R}$ for $j = 1, 2, \dots, p$ by

$$y_j(t) = \int_0^\infty y(t-s)g_a^j(s)ds = \int_{-\infty}^t y(s)g_a^j(t-s)ds \tag{7.12}$$

Then for $j > 1$, where $g_a^j(0) = 0$, we have

$$\begin{aligned}
y_j'(t) &= \frac{d}{dt}[\int_{-\infty}^t y(\eta)g_a^j(t-\eta)d\eta] \\
&= y(t)g_a^j(0) + \int_{-\infty}^t y(\eta)\frac{d}{dt}g_a^j(t-\eta)d\eta \\
&= \int_{-\infty}^t y(\eta)a[g_a^{j-1}(t-\eta) - g_a^j(t-\eta)]d\eta \\
&= \int_0^\infty y(t-s)a[g_a^{j-1}(s) - g_a^j(s)]ds \\
&= a[y_{j-1}(t) - y_j(t)]
\end{aligned}$$

whereas for $j = 1$, where $g_a^1(0) = a$, the same argument gives

$$y_1'(t) = a[y(t) - y_1(t)]$$

Letting $y_0(t) = y(t)$, we see that the $y_j(t)$ satisfy the nonhomogeneous linear system of ODEs

$$y_j'(t) = a[y_{j-1}(t) - y_j(t)], j = 1, 2, \dots, p, \tag{7.13}$$

on $(-\infty, T)$.

It turns out that if $T = \infty$, then (7.12) gives the unique solution of (7.13) that is bounded on $\mathbb{R}$.

Proposition 7.3 *For each bounded continuous function* $y_0 : \mathbb{R} \to \mathbb{R}$, *there is a unique solution* $y = (y_1, \dots, y_p)^T$ *of*

$$y_j'(t) = a[y_{j-1}(t) - y_j(t)], \ j = 1, 2, \dots, p \tag{7.14}$$

that is bounded on $\mathbb{R}$ *and it is given by*

$$y_j(t) = \int_0^\infty y_0(\eta)g_a^j(t-\eta)d\eta, j = 1, 2, \dots, p. \tag{7.15}$$

Proof. Vector function y satisfies

$$y' = Ay + ay_0(t)e_1$$

where A is the $p \times p$ matrix with $-a$ on main diagonal, a on subdiagonal and all other entries zero; $e_1 = (1, 0, \ldots, 0)^T$. It is easily seen that the homogeneous system has only the trivial solution that is bounded on $\mathbb{R}$. Consequently, (7.14) can have at most one bounded solution because the difference of two bounded solutions gives a bounded solution of the homogeneous equation. The variation of constants formula gives

$$y(t) = e^{A(t-s)}y(s) + \int_s^t e^{A(t-\eta)}ay_0(\eta)e_1 d\eta$$

and letting $s \to -\infty$ leads to an expression for the unique bounded solution on $\mathbb{R}$

$$y(t) = \int_{-\infty}^t ay_0(\eta)e^{A(t-\eta)}e_1 d\eta$$

See Theorem IV.1.1 of [40]. $v(t) = e^{At}e_1$ satisfies $v_1' = -av_1, v_1(0) = 1$, and $v_j' = a[v_{j-1} - v_j]$, $v_j(0) = 0$. Clearly, $av(t)$ satisfies (7.11). $\square$

Consider the nonlinear equation

$$y'(t) = F\left(\int_0^\infty y(t-s)g_a^p(s)ds \right) \tag{7.16}$$

where $F : \mathbb{R} \to \mathbb{R}$ is continuously differentiable, $F(0) = 0$, and $F'(0) = -k$ with $k > 0$. Of course, we must prescribe initial conditions:

$$y(\theta) = \phi(\theta), \theta \le 0,$$

where $\phi : (-\infty, 0] \to \mathbb{R}$ is bounded and continuous. Suppose that $T > 0$ and $y : (-\infty, T) \to \mathbb{R}$ is a solution of this initial-value problem. By this, we mean that y is continuous and differentiable on $[0, T)$ and satisfies (7.16) there. Let y_j, $j = 1, 2, \ldots, p$ be defined by (7.12). Then it follows that, with $y_0(t) = y(t)$, we have

$$y_0'(t) = F(y_p(t)), 0 \le t < T \tag{7.17}$$
$$y_j'(t) = a[y_{j-1}(t) - y_j(t)]$$

with initial data

$$y_0(0) = \phi(0)$$
$$y_j(0) = \int_0^\infty \phi(-s)g_a^j(s)ds, j = 1, 2, \ldots, p$$

The procedure by which we obtain an ODE such as (7.17) from a delay system such as (7.16) is referred to as the "linear chain trick".

Of course, it is important to be able to go in the other direction as well. If one finds an interesting solution of the ODEs (7.17), for example, a periodic solution, then one wants to know that it is a solution of (7.16). Proposition 7.3 allows us to conclude that any solution of (7.17) which is bounded on the entire real line is also a solution of the infinite delay equation (7.16).

Observe that the linearization of (7.16) about the trivial solution is given by (7.7) with characteristic equation (7.10). But (7.10) is also the characteristic equation for the eigenvalues of the linearization of system (7.17) about the trivial solution. As observed in Remark 7.2, purely imaginary roots occur under suitable conditions, so we may expect the possibility of a Hopf bifurcation of periodic solutions of (7.17) from the trivial solution. Of course, we must identify a bifurcation parameter. In any case, should periodic solutions of (7.17) exist, they would also be solutions of (7.16).

In applications, one may seek to avoid the complications of using delay differential equations by simply utilizing the ODE system to represent a time delay. In this case, one uses (7.4) to determine appropriate a, p.

7.2 A Model of HIV Transmission

Culshaw, Ruan, and Webb [23] cite evidence that in lymphatic tissue direct cell-to-cell transmission of HIV is the dominant mode of infection. If C denotes concentration of healthy cells and I is concentration of infected cells, they derive the model system:

$$C'(t) = rC(t)\left(1 - \frac{C(t) + I(t)}{C_M}\right) - k_I I(t)C(t)$$

$$I'(t) = k_I' \int_{-\infty}^{t} I(u)C(u)g_a^p(t - u)du - \mu_I I(t) \qquad (7.18)$$

k_I'/k_I is the fraction of cells surviving the incubation period.

Equilibria of (7.18) are nonnegative solutions of

$$0 = rC\left(1 - \frac{C + I}{C_M}\right) - k_I IC$$

$$0 = k_I' IC - \mu_I I$$

They consist of the disease-free equilibrium $(C, I) = (C_M, 0)$ and the disease equilibrium

$$(\bar{C}, \bar{I}) = (\mu_I/k_I', \ r(1 - \mu_I/k_I' C_M)(k_I + 1/C_M)^{-1})$$

which is nonnegative only when $\mu_I < k_I' C_M$.

The linearized system about the disease-free equilibrium is given by

$$u'(t) = -ru(t) - (k_I C_M + r)v(t)$$

$$v'(t) = C_M k_I' \int_0^\infty v(t - s)g_a^p(s)ds - \mu_I v(t)$$

$(u, v)^T = e^{\lambda t}(c, d)^T$ is a solution if

$$\begin{pmatrix} -r-\lambda & -(k_I C_M + r) \\ 0 & \frac{a^p C_M k'_I}{(a+\lambda)^p} - \lambda - \mu_I \end{pmatrix} \begin{pmatrix} c \\ d \end{pmatrix} = \begin{pmatrix} 0 \\ 0 \end{pmatrix}$$

and it is nonzero if the determinant of the matrix vanishes. This gives the characteristic equation

$$(-r-\lambda)\left(\frac{a^p C_M k'_I}{(a+\lambda)^p} - \lambda - \mu_I\right) = 0$$

The first factor gives $\lambda = -r$ so the second decides stability:

$$0 = \frac{a^p C_M k'_I}{(a+\lambda)^p} - \lambda - \mu_I \tag{7.19}$$

By equating the first term on the right with $\lambda + \mu_I$ we see from a graphical analysis that there is precisely one real root larger than $-a$ and this root is positive if and only if $C_M k'_I > \mu_I$. This inequality implies instability of the disease-free equilibrium but it may not be sharp because it ignores complex roots.

Suppose there is a root λ with nonnegative real part. Then

$$\begin{aligned} a^p C_M k'_I &= (a+\lambda)^p(\lambda + \mu_I) \\ &= |(a+\lambda)^p(\lambda+\mu_I)| = |a+\lambda|^p|\lambda+\mu_I| \\ &\geq a^p \mu_I \end{aligned}$$

where the final inequality follows because $|a+\lambda| \geq a$ and $|\lambda + \mu_I| \geq \mu_I$ for $\Re\lambda \geq 0$. So $C_M k'_I \geq \mu_I$. It follows that all roots have negative real part if $C_M k'_I < \mu_I$. Therefore, the disease-free equilibrium is asymptotically stable when $C_M k'_I < \mu_I$ and unstable when the reverse inequality holds.

In summary, when $C_M k'_I < \mu_I$ the disease-free equilibrium is the only equilibrium and it is asymptotically stable; with the inequality reversed, the disease-free equilibrium is unstable and the disease equilibrium exists.

The linearized system about the disease equilibrium $(\bar{C}, \bar{I})$ is given by

$$u'(t) = -\alpha u(t) - \beta v(t)$$
$$v'(t) = \int_0^\infty [\mu_I v(t-s) + k'_I \bar{I} u(t-s)]g_a^p(s)ds - \mu_I v(t)$$

where

$$\alpha = \frac{r\mu_I}{k'_I C_M}, \beta = (k_I + r/C_M)\frac{\mu_I}{k'_I}$$

It has an exponential solution $(u,v)^T = e^{\lambda t}(c,d)^T$ if

$$\begin{pmatrix} -\alpha - \lambda & -\beta \\ k'_I \bar{I} f & \mu_I f - \lambda - \mu_I \end{pmatrix} \begin{pmatrix} c \\ d \end{pmatrix} = \begin{pmatrix} 0 \\ 0 \end{pmatrix}$$

where $f = \int_0^\infty e^{-\lambda u} g_a^p(u)du = a^p/(a+\lambda)^p$. The resulting characteristic equation is given by

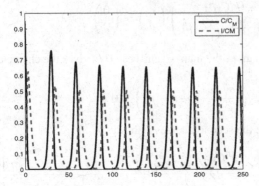

Fig. 7.2 Simulation of (7.21) with $p = 2$ and $a = 2$.

$$0 = (\lambda + \alpha)(\lambda + \mu_I - \mu_I \frac{a^p}{(a+\lambda)^p}) + \beta k_I' \bar{I} \frac{a^p}{(a+\lambda)^p} \qquad (7.20)$$

Observe that the right side is positive for all $\lambda \geq 0$ so any eigenvalues with nonnegative real part must have nontrivial imaginary parts.

Culshaw et al. [23] show that when $p = 1$, the disease equilibrium is asymptotically stable for small values of a and their analytical results suggest the possibility of a Hopf bifurcation at a critical value of a. Their simulations for the case $p = 1$ support these assertions. In particular, their numerical simulations of the associated ODE model (see below) suggest the existence of an orbitally asymptotically stable periodic solution for suitable parameter values. According to Proposition 7.3, any periodic solution for the ODE is also a solution of (7.18).

The family of ODE systems with $p = 1, 2, \ldots$ obtained from (7.18) by the linear chain trick

$$y_j(t) = \int_{-\infty}^{t} I(u)C(u)g_a^j(t-u)du, j = 1, 2, \ldots p$$

is given by

$$\begin{aligned}
C'(t) &= rC(t)\left(1 - \frac{C(t)+I(t)}{C_M}\right) - k_I I(t)C(t) \\
y_1'(t) &= a[I(t)C(t) - y_1(t)] \\
y_j'(t) &= a[y_{j-1}(t) - y_j(t)], 2 \leq j \leq p \\
I'(t) &= k_I' y_p(t) - \mu_I I(t)
\end{aligned} \qquad (7.21)$$

Evidence for the existence of periodic solutions of (7.21) with $p = 2$ and $a = 2$ is provided in the numerical simulation depicted in Figure 7.2. Equation (7.21) was scaled for the simulation by setting:

$$\bar{C} = C/C_M, \bar{I} = I/C_M, \bar{y}_i = y_i/C_M^2.$$

Parameters used in Figure 7.2 are $k_I C_M = 4$, $k'_I C_M = 3.5$, $r = 0.68$, $\mu_I = 0.3$.

7.3 An ODE Approximation to a Delay Equation

Consider the discrete delay equation

$$x'(t) = f(x(t), x(t-1)) \tag{7.22}$$

Let us assume that $x(t)$ is a solution with initial data given by $x_0 = \phi$. We want to approximate $x(t)$.

Fix a positive integer N and let

$$x_i(t) = x\left(t - \frac{i}{N}\right), 0 \le i \le N$$

Note that $x_0(t) = x(t)$, $x_N(t) = x(t-1)$, and

$$x'_i(t) \approx N[x_i(t+1/N) - x_i(t)] = N[x_{i-1}(t) - x_i(t)], t > i\frac{1}{N}$$

The delay differential equation can be approximated by the system of $N+1$ ODEs

$$\begin{aligned} x'_0(t) &= f(x_0(t), x_N(t)) \\ x'_i(t) &= N[x_{i-1}(t) - x_i(t)], 1 \le i \le N \end{aligned} \tag{7.23}$$

where $x_i(0) = \phi(-i/N)$. There is a considerable literature on this approximation technique. Gedeon and Hines [35] relate the global attractor of the delay equation to the global attractor of the discretization. W. Stone uses the above approximation in his PhD thesis under Frank Hoppensteadt's direction [74] in order to study nonlinear oscillations in Wright's equation.

Exercises

Exercise 7.1. Verify that

$$F_p(\lambda) \equiv \int_0^\infty e^{-\lambda s} \frac{a^p s^{p-1} e^{-as}}{(p-1)!} ds = \frac{a^p}{(a+\lambda)^p}, \Re\lambda > -a$$

by showing that

$$\frac{d}{d\lambda} F_p(\lambda) = -\frac{p}{a} F_{p+1}(\lambda)$$

and then calculating $F_1(\lambda) = a/(a+\lambda)$.

Exercise 7.2. Solve (7.9) where $k(s) = -g_1^2(s)$ and $\phi(t) = e^{-t}$. Use the Laplace transform.

Exercise 7.3. Verify the assertions of Proposition 7.1 concerning the real roots of (7.10).

Exercise 7.4. Show that by the choice $p = n$, $a = n/r$, and taking the limit $n \to \infty$ in the characteristic equation (7.10), we get the characteristic equation

$$\lambda = ke^{-\lambda r}$$

Exercise 7.5. Verify that the g_a^j, $j = 1, 2, \ldots, p$ satisfy (7.11).

Exercise 7.6. Consider the delayed logistic equation

$$N'(t) = N(t)[b - cN(t) - d \int_{-\infty}^{t} N(s)g_a^2(t-s)ds]$$

where $b, d > 0$, $c \geq 0$. Thus, $T = a/2$ is the "average delay." See [24] for more on such equations.

(a) Determine the stability of $N = 0$.
(b) Find the characteristic equation corresponding to the positive steady state.
(c) If $c = 0$, determine conditions on the average delay T such that the positive
    steady state is asymptotically stable.
(d) Apply the linear chain trick to obtain the ODE associated with this equation.
(e) Can a Hopf bifurcation occur from the positive steady state?

Exercise 7.7. Apply the linear chain trick to obtain the ODEs associated with the linear system (7.6). Show that it takes the form $y' = Ay$ where $(p+1) \times (p+1)$ matrix A is as follows. The only nonzero entry of the first row is $-k$ in the last column; the nonzero entries in row $(i+1)$ are a in the ith column and $-a$ in the $i+1$-column (the diagonal entry), $i = 1, 2, \ldots, p$. If $k, a > 0$, obtain necessary and sufficient conditions for all eigenvalues to have negative real part.

Exercise 7.8. Verify the linearizations of (7.18) and associated characteristic equations about the disease-free and the disease equilibrium.

Exercise 7.9. Consider the equation (7.22) where $f(x,y) = -\beta y$ and the ODE approximation (7.23). Relate the eigenvalue of the Jacobian of (7.23) at $x = 0$ with the maximal real part with the corresponding root of the characteristic equation (2.10).

Chapter 8
Phage and Bacteria in a Chemostat

Abstract A mathematical model of bacteriophage predation on bacteria in a chemostat, introduced in Chapter 1, is studied in detail. The delay arises due to the assumption of a fixed latent period for virus inside the infected cell. The main biological issues addressed are the persistence and extinction of bacteria and of bacteriophage (phage). From a mathematical perspective, we focus on bifurcation of equilibria and Hopf bifurcation from the coexistence equilibrium.

8.1 Model

Levin, Stewart, and Chao [50] and Lenski and Levin [51] model phage (virus that attack bacteria) predation on a bacterial host which in turn consumes a limiting nutrient in a chemostat by the system

$$
\begin{aligned}
R'(t) &= D(R_0 - R(t)) - f(R(t))S(t) \\
S'(t) &= (f(R(t)) - D)S(t) - kS(t)P(t) \\
I'(t) &= kS(t)P(t) - DI(t) - e^{-D\tau}kS(t-\tau)P(t-\tau) \\
P'(t) &= -DP(t) - kS(t)P(t) + be^{-D\tau}kS(t-\tau)P(t-\tau)
\end{aligned}
\tag{8.1}
$$

R is the resource supporting bacterial growth, S is uninfected bacteria, I is phage-infected bacteria, and P is phage. A recent mathematical analysis of the model was carried out by Beretta, Solimano, and Tang [7]. R_0 is input nutrient concentration supplied to bacteria, D is the dilution rate of the chemostat, and $f(R)$ is the specific growth rate of bacteria at resource level R. The specific growth rate f is typically taken to be of Monod type:

$$
f(R) = \frac{mR}{a+R}
$$

where $m, a > 0$. However, we need only assume that $f : \mathbb{R}_+ \to \mathbb{R}_+$ is C^1 and

H. Smith, *An Introduction to Delay Differential Equations with Applications to the Life Sciences,*   131
Texts in Applied Mathematics 57, DOI 10.1007/978-1-4419-7646-8_8,
© Springer Science+Business Media, LLC 2011

$$f(0) = 0, f'(R) > 0, f(\infty) < \infty. \tag{8.2}$$

Phage attach to the cell surface of a bacterium and inject their DNA into it. This causes the bacterium to begin to synthesize viral DNA and viral proteins in order to make new virus. After a time τ, called the latent period, this is complete and the bacterium lyses open releasing the new virus. Latent periods vary by bacterial type but are usually in the half hour to hour range. Denote by $b \geq 1$ the average number of progeny released when an infected cell lyses. The factor $e^{-D\tau}$ in the equations accounts for the fraction of infected bacteria that survive being washed out of the chemostat during the latent period. More generally, the probability of phage, nutrient, or bacteria avoiding washout in a time period of length t is e^{-Dt}.

Two important assumptions are made in formulating the model: (1) nutrient uptake by infected cells is negligible, and (2) infected cells do not grow and divide. Phage binding to infected cells has also been neglected. We have scaled out the yield constant, a positive number multiplying $f(R)$ in the equation for R.

Nonnegative initial data must be prescribed for (8.1). Clearly, S and P must be given on $[-\tau, 0]$ but only $R(0)$ and $I(0)$ are needed. However, it is easy to see that $I(0)$ cannot be given any nonnegative value, independent of those of S and P. For example, if $I(0) = 0$ but $e^{-D\tau}kS(-\tau)P(-\tau) > kS(0)P(0)$, then $I'(0) < 0$ so I becomes negative. It is reasonable to expect that $I(t)$ can be obtained directly from knowledge of $S(\eta)$ and $P(\eta)$ for $\eta \in [t-\tau, t]$. In fact, the formula

$$I(t) = \int_{t-\tau}^{t} ke^{-D(t-\eta)} S(\eta)P(\eta)d\eta = \int_{0}^{\tau} e^{-Ds}kS(t-s)P(t-s)ds \tag{8.3}$$

makes good biological sense because for $\eta \in [t-\tau, t]$, $kS(\eta)P(\eta)d\eta$ represents the new infections occurring in the interval $[\eta, \eta + d\eta]$, and $e^{-D(t-\eta)}$ represents the fraction of these that are still in the chemostat at time t. These cells have not yet been infected for τ units of time and so they have not yet lysed. On the other hand, it is easy to verify that (8.3) satisfies the differential equation for I in (8.1) and the initial condition

$$I(0) = \int_{0}^{\tau} e^{-Ds}kS(-s)P(-s)ds.$$

Busenberg and Cooke [13] were among the first to point out these restrictions on initial data for equations like (8.1).

The bottom line is that I is given by (8.3) so we may consider the R, S, P system:

$$\begin{aligned} R'(t) &= D(R_0 - R(t)) - f(R(t))S(t) \\ S'(t) &= (f(R(t)) - D)S(t) - kS(t)P(t) \\ P'(t) &= -DP(t) - kS(t)P(t) + be^{-D\tau}kS(t-\tau)P(t-\tau) \end{aligned} \tag{8.4}$$

Nonnegative initial data for S and P must be prescribed on $[-\tau, 0]$ but only $R(0)$ need be prescribed:

$$R(0) = R(0) \tag{8.5}$$
$$S(s) = \phi(s), s \in [-\tau, 0]$$
$$P(s) = \psi(s), s \in [-\tau, 0]$$

In this formulation, the state of the system at time t is the triple $(R(t), S_t, P_t) \in \mathbb{R}_+ \times C([-\tau, 0], \mathbb{R}_+)^2$.

8.2 Positivity and Boundedness of Solutions

The state space for system (8.4) is

$$X = \mathbb{R}_+ \times C([-\tau, 0], \mathbb{R}_+) \times C([-\tau, 0], \mathbb{R}_+) \tag{8.6}$$

The next result ensures that solutions exist and are positive and bounded.

Theorem 8.1 *The initial-value problem* (8.4) *with nonnegative initial data* (8.5) *has a unique nonnegative solution* $(R(t), S(t), P(t))$ *defined for all* $t \geq 0$ *which is bounded. In fact, the following estimate holds* $Q(t) = R(t) + S(t) + I(t) + P(t)/b$:

$$Q(t) \leq \max\{Q(0), R_0\} \tag{8.7}$$

Equations (8.4) *and* (8.5) *induce a continuous semiflow* $\Phi : \mathbb{R}_+ \times X \to X$ *defined by*

$$\Phi(t, (R(0), \phi, \psi)) = (R(t), S_t, P_t)$$

Proof. Local existence follows from Theorem 3.1 and Remark 3.3. Positivity of solutions is implied by Theorem 3.4. Continuation of our nonnegative solution to $t \geq 0$ requires an a priori bound (Theorem 3.2). Observe that

$$\left(R + S + I + \frac{1}{b}P\right)' \leq D[R_0 - (R + S + I + \frac{1}{b}P)]$$

which implies that

$$Q(t) \leq Q(0)e^{-Dt} + R_0(1 - e^{-Dt})$$

and hence

$$\limsup_{t \to \infty} Q(t) \leq R_0$$

and (8.7) holds. This shows that solutions exist for all $t \geq 0$ and are bounded. $\square$

The following proposition proves useful later on. It describes initial data that result in no phage or infected cells.

Proposition 8.2 *If* $P(0) = \psi(0) = 0$ *and* $I(0) = \int_0^\tau e^{-Ds} k\phi(-s)\psi(-s)ds = 0$, *then* $P(t) = I(t) = 0$, $t \geq 0$.

Proof. $u = I + P$ satisfies $u(t) \geq 0$ and $u' = -Du + (b-1)e^{-D\tau}kS(t-\tau)P(t-\tau)$ on $t \geq 0$ and $u(0) = 0$. As $I(0) = 0$, it follows that $\phi(-s)\psi(-s) = 0$, $0 \leq s \leq \tau$ so $S(t-\tau)P(t-\tau) = 0$ for $0 \leq t \leq \tau$ and consequently $u(t) = 0$ on this interval. But then $S(t-\tau)P(t-\tau) = 0$ on the larger interval $0 \leq t \leq 2\tau$ and we can repeat the argument by taking steps of length τ. $\square$

8.3 Basic Reproductive Number for Phage

We assume that susceptible bacteria are viable in the absence of phage. By this we mean that the phage-free system

$$R'(t) = D(R_0 - R(t)) - f(R(t))S(t)$$
$$S'(t) = (f(R(t)) - D)S(t) \tag{8.8}$$

has a unique positive equilibrium $(\bar{R}, \bar{S})$

$$\bar{R} = f^{-1}(D), \bar{S} = R_0 - \bar{R} \tag{8.9}$$

This is easily seen to be the case if and only if

$$f(R_0) > D \tag{8.10}$$

It is well known that this equilibrium attracts all solutions of (8.8) with $S(0) > 0$. Of course, (8.8) contains no delay terms so appropriate initial data are nonnegative values for $R(0), S(0)$.

System (8.4) has two "boundary equilibria," namely

$$E_R = (R_0, 0, 0) \quad \text{and} \quad E_S = (\bar{R}, \bar{S}, 0)$$

Proposition 8.3 *Let (8.10) hold. Then E_S is locally asymptotically stable for (8.4) if the phage reproductive number*

$$PRN \equiv \frac{be^{-D\tau}k\bar{S}}{D + k\bar{S}} \tag{8.11}$$

satisfies $PRN < 1$ and unstable if $PRN > 1$.

Proof. The linearization of (8.4) about the phage-free equilibrium $E_S = (\bar{R}, \bar{S}, 0)$ of (8.4) can be written as

$$x'(t) = Ax(t) + Bx(t-\tau)$$

where $x(t) = (R(t), S(t), P(t))^T$ and

$$A = \begin{pmatrix} a_{11} & a_{12} & 0 \\ a_{21} & 0 & a_{23} \\ 0 & 0 & -(D+k\bar{S}) \end{pmatrix} \quad \text{and} \quad B = \begin{pmatrix} 0 & 0 & 0 \\ 0 & 0 & 0 \\ 0 & 0 & be^{-D\tau}k\bar{S} \end{pmatrix}$$

where
$$a_{11} = -D - f'(\bar{R})\bar{S}, a_{12} = -D, a_{21} = f'(\bar{R})\bar{S}, a_{23} = -k\bar{S}$$

The characteristic equation factors nicely as

$$(\lambda^2 - a_{11}\lambda - a_{12}a_{21})(\lambda + D + k\bar{S} - be^{-D\tau}k\bar{S}e^{-\lambda\tau}) \qquad (8.12)$$

The roots of the quadratic have negative real part as expected because $(\bar{R}, \bar{S})$ is asymptotically stable in the linear approximation for system (8.8). Therefore, stability boils down to the factor:

$$\lambda = -(D + k\bar{S}) + be^{-D\tau}k\bar{S}e^{-\lambda\tau} \qquad (8.13)$$

Using Theorem 4.7 and Exercise 4.9 and the notation used therein, namely $A = -(D + k\bar{S})$ and $B = be^{-D\tau}k\bar{S}$, we have $A + B = -D + k\bar{S}(-1 + be^{-D\tau})$ and $B \geq A$. Hence, all roots have negative real part if $A + B < 0$ and there is a positive root if $A + B > 0$. But $A + B > 0$ if and only if $PRN > 1$; $A + B < 0$ if and only if $PRN < 1$. The result follows from Theorem 4.3. $\square$

Remark 8.4 *PRN has a simple biological interpretation. Imagine adding a single hypothetical phage to the chemostat at the phage-free equilibrium E_S. Two possibilities can occur, the phage can wash out before it can bind and infect a bacterium or it can bind and infect a bacterium. The probability of the latter is $k\bar{S}/(D + k\bar{S})$. Assuming the phage binds, then the same two possibilities apply to the resulting infected cell. The probability that it remains in the chemostat through the latent period is $e^{-D\tau}$. It then lyses releasing b progeny phage. Therefore, the expected value of the number of progeny resulting from adding a single phage to the chemostat at equilibrium E_S is PRN. It may be called the phage reproductive number or ratio.*

8.4 Persistence of Host and Phage Extinction

If $f : [0, \infty) \to \mathbb{R}$ we write $f_\infty = \liminf_{t\to\infty} f(t)$ and $f^\infty = \limsup_{t\to\infty} f(t)$. We show below that the bacteria cannot be driven to extinction by the phage if (8.10) holds.

Proposition 8.5 *Let (8.10) hold. If $(R(t), S(t), P(t))$ is a positive solution of (8.4), then*

$$\bar{R} \leq R_\infty \leq R^\infty \leq R_0 \quad and \quad 0 < S^\infty \leq \bar{S}$$

Proof. As $R' \leq D(R_0 - R)$ we conclude that $R^\infty \leq R_0$.

We claim that $S^\infty > 0$. Otherwise, $S(t) \to 0$. Then from the R equation, we see that $R(t) \to R_0$. Now apply the fluctuation Lemma A.1 to the equation for P with $P(s_n) \to P^\infty$, and $P'(s_n) \to 0$ to find that $P^\infty = 0$ so $P(t) \to 0$. But then, by (8.10), there exists $\varepsilon > 0$ such that $f(R(t)) - D - kP(t) > \varepsilon$ for large t so $S'(t) \geq \varepsilon S(t)$ for such t. This gives the contradiction $S^\infty = \infty$. We conclude that $S^\infty > 0$.

Write $M = R + S$ and the equations for M, S become

$$M' = D(R_0 - M) - kSP$$
$$S' = (f(M - S) - D)S - kSP$$

Clearly, $M^\infty \le R_0$ follows from the first equation. By the fluctuation Lemma A.1, there exists $t_n \to 0$ such that $S(t_n) \to S^\infty > 0$, $S'(t_n) \to 0$ and we may assume that $P(t_n) \to P$ and $M(t_n) \to M$. As $S^\infty > 0$, the equation for S gives

$$0 = f(M - S^\infty) - D - kP \le f(R_0 - S^\infty) - D$$

Because $f(\bar{R}) - D = 0$, we conclude that $\bar{R} \le R_0 - S^\infty$ or that $S^\infty \le \bar{S} = R_0 - \bar{R}$.

By the fluctuation Lemma A.1, there exists $t_n \to 0$ such that $R(t_n) \to R_\infty$, $R'(t_n) \to 0$ and we may assume that $S(t_n) \to S$. From the paragraph above, $S \le \bar{S}$. From the differential equation for R we find

$$0 = D(R_0 - R_\infty) - f(R_\infty)S \ge D(R_0 - R_\infty) - f(R_\infty)\bar{S}$$

The right-hand side vanishes if R_∞ is replaced by $\bar{R}$. It follows from monotonicity of $R \to D(R_0 - R) - f(R)\bar{S}$ that $\bar{R} \le R_\infty$.   □

Following Beretta, Solimano, and Tang [7], we have the following result which implies phage extinction when $PRN < 1$ and Monod uptake function f.

Proposition 8.6 *If $PRN < 1$ and*

$$c = \frac{f(R)(R - \bar{R})}{R(f(R) - f(\bar{R}))} \tag{8.14}$$

is independent of R, then $(R(t), S(t), P(t)) \to E_S$ for every solution with $S(0) > 0$.

Proof. Consider the Liapunov function

$$V(t) = R(t) - \bar{R}\ln R(t) + c_1(S(t) - \bar{S}\ln S(t)) + c_2(bI(t) + P(t))$$

where $c_1 > 0$ and $c_2 > 0$ must be determined. We regard I as given by (8.3) in terms of S and P. Then, using that $D(R_0 - \bar{R}) - f(\bar{R})\bar{S} = 0$ and $f(\bar{R}) = D$, we find that

$$V' = \frac{R - \bar{R}}{R}R'(t) + c_1\frac{S - \bar{S}}{S}S'(t) + c_2(bI'(t) + P'(t))$$

$$= \frac{R - \bar{R}}{R}(D(R_0 - R) - f(R)S) + c_1(S - \bar{S})(f(R) - D - kP) + c_2((b - 1)kSP$$
$$- D(bI + P))$$

$$= -\frac{R - \bar{R}}{R}\left(D(R - \bar{R}) + D(S - \bar{S}) + (f(R) - f(\bar{R}))S\right) + c_1(S - \bar{S})(f(R) - f(\bar{R}))$$
$$- c_1kSP + c_1\bar{S}kP + c_2((b - 1)kSP - D(bI + P))$$

$$= -D\frac{(R - \bar{R})^2}{R} + (R - \bar{R})(S - \bar{S})\left(c_1\frac{f(R) - f(\bar{R})}{R - \bar{R}} - \frac{(f(R)S - f(\bar{R})\bar{S})}{R(S - \bar{S})}\right)$$
$$- c_2\left(DbI + (D - (b - 1)k\bar{S})P\right)$$

provided $c_1 = (b-1)c_2$. Note that $q \equiv D - (b-1)k\bar{S} > 0$ inasmuch as $PRN < 1$.

Now take $c_1 = c$ given by (8.14) and note that

$$c_1 \frac{f(R) - f(\bar{R})}{R - \bar{R}} = \frac{f(R)}{R}$$

This leads to

$$V' = -D\frac{(R-\bar{R})^2}{R} - \frac{(f(R) - f(\bar{R}))(R-\bar{R})\bar{S}}{R} - c_2(DbI + qP) \leq 0. \qquad (8.15)$$

Because we wish to apply Theorem 5.17 to obtain the desired conclusion, it is necessary to formally define $V : X_0 \to \mathbb{R}$ on a subset $X_0 \subset X$. Let $X_0 = \{(R, \phi, \psi) \in X : R(0) > 0, \phi(0) > 0\}$ and observe that X_0 is positively invariant for (8.8). Then $V = V(R, \phi, \psi)$ is given on X_0 by

$$V = R - \bar{R}\ln R + c_1\left(\phi(0) - \bar{S}\ln\phi(0)\right) + c_2\left(b\int_0^\tau e^{-Ds}k\phi(-s)\psi(-s)ds + \psi(0)\right)$$

It is easy to see that V is continuous on X_0.

Given $(R, \phi, \psi) \in X_0$, define $G \subset X_0$ by

$$G = \{(\hat{R}, \hat{\phi}, \hat{\psi}) \in X_0 : V(\hat{R}, \hat{\phi}, \hat{\psi}) \leq V(R, \phi, \psi)\}$$

By continuity of V, G is closed, and our calculation above shows that G is positively invariant for (8.8). The omega limit set of the orbit through (R, ϕ, ψ), denoted by ω, must therefore satisfy $\omega \subset G$. In fact, by Theorem 5.17, ω must belong to the largest invariant subset S of $M \equiv \{(\hat{R}, \hat{\phi}, \hat{\psi}) \in G : V' = 0\}$. By (8.15), $(\hat{R}, \hat{\phi}, \hat{\psi}) \in M$ if and only if $\hat{R} = \bar{R}$, $\hat{\psi}(0) = 0$, and

$$I = \int_0^\tau e^{-Ds}k\hat{\phi}(-s)\hat{\psi}(-s)ds = 0.$$

Hence, by Proposition 8.2, the solution $(\hat{R}(t), \hat{S}(t), \hat{P}(t))$ of (8.8) with initial data $(\hat{R}, \hat{\phi}, \hat{\psi}) \in M$ satisfies $\hat{P}(t) = 0$, $t \geq 0$. This must hold for every solution through a point of ω and so ω must be an invariant set of the chemostat system (8.8) with $R = \bar{R}$. The only such point is the equilibrium E_S. Thus ω consists of the equilibrium state corresponding to E_S. $\square$

The hypothesis (8.14) holds in the case where f is of Monod type $f(R) = mR/(a+R)$.

For more on persistence of phage, see [72].

8.5 The Coexistence Equilibrium

System (8.4) may have at most one positive equilibrium

$$E_C = (R^*, S^*, P^*)$$

corresponding to the coexistence of phage and bacteria. It is determined by the relations:

$$S^* = \frac{D}{k(be^{-D\tau} - 1)}, S^* = \frac{D(R_0 - R^*)}{f(R^*)}, kP^* = f(R^*) - D \qquad (8.16)$$

The equation on the left determines S^*, the middle one determines (implicitly) R^*, from which the final equation gives P^*. Observe that unlike equilibria E_R and E_S, the coordinates of E_C depend on the delay τ. The manner in which E_C depends on τ is described below.

Lemma 8.1. E_C *exists with all positive components if and only if* (8.10) *and PRN* $>$ 1 *hold. If it exists, then* $\bar{R} < R^* < R_0$ *and* $S^* < \bar{S}$.
$PRN > 1$ *if and only if* $bk\bar{S}/(D + k\bar{S}) > 1$.
If $bk\bar{S}/(D + k\bar{S}) > 1$ *and* (8.10) *hold then* E_C *exists for all values of the delay* τ *satisfying* $0 \leq \tau < \tau_c$, *where* $0 < \tau_c < \infty$ *is characterized by the equation*

$$PRN = \frac{be^{-D\tau_c} k\bar{S}}{D + k\bar{S}} = 1$$

Moreover, for $0 \leq \tau < \tau_c$, $R^*(\tau)$ *and* $P^*(\tau)$ *are decreasing and* $S^*(\tau)$ *is increasing in* τ *and*

$$E_C(\tau) \to E_S, \tau \nearrow \tau_c$$

Existence and positivity of E_C *arise through a transcritical bifurcation from* E_S *as delay* τ *passes below* τ_c.

Proof. Suppose first that E_C exists. From (8.16) we have $\bar{R} < R^* < R_0$. Monotonicity of $R \to D(R_0 - R)/f(R)$ on $(0, R_0)$ implies that $S^* < \bar{S}$. Therefore, $\bar{S} > S^* = D/(k(be^{-D\tau} - 1))$, implying that $PRN > 1$.

Conversely, if (8.10) and $PRN > 1$ hold then $k\bar{S}(be^{-D\tau} - 1) > D$ so there is a unique value of $S^* \in (0, \bar{S})$ satisfying $kS^*(be^{-D\tau} - 1) = D$. But then $R^* \in (\bar{R}, R_0)$ is uniquely defined by the relation in (8.16) due to monotonicity of $R \to D(R_0 - R)/f(R)$ and similarly for P^*.

The monotonicity assertions concerning R^*, S^*, P^* in the variable τ follow directly from the relations (8.16) and the fact that f is increasing and $R \to D(R_0 - R)/f(R)$ is decreasing on $0 < R \leq R_0$. At $\tau = \tau_c$, $(be^{-D\tau_c} - 1)k\bar{S} = D$ so $S^* = \bar{S}$ and this implies $R^* = \bar{R}$ so $f(R^*) = D$ and hence $P^* = 0$.

If $PRN > 1$ and (8.10) hold, then we can compute the derivatives $M_\tau^* = (dM^*/d\tau)(\tau_c)$, $M = R, S, P$. The results of a tedious implicit differentiation are:

$$R_\tau^* = -\frac{bkD\bar{S}^2}{D + f'(\bar{R})\bar{S}} e^{-D\tau_c}$$

$$S_\tau^* = bk\bar{S}^2 e^{-D\tau_c} \qquad (8.17)$$

$$P_\tau^* = -\frac{bDf'(\bar{R})\bar{S}^2}{D + f'(\bar{R})\bar{S}} e^{-D\tau_c}$$

Observe that $P(\tau_c) = 0$ and $P_\tau^*(\tau_c) < 0$. This establishes the transcritical bifurcation.
□

The central question is whether the phage and bacteria can coexist. Our aim here is to show that the phage P and bacteria S can stably coexist if (8.10) and $PRN > 1$ and additional conditions hold. We now proceed to determine the stability of E_C.

The linearization about E_C is determined by the matrices:

$$A = A(\tau) = \begin{pmatrix} -D - S^* f'(R^*) & -f(R^*) & 0 \\ S^* f'(R^*) & 0 & -kS^* \\ 0 & -kP^* & -kS^* - D \end{pmatrix}$$

and

$$B = B(\tau) = \begin{pmatrix} 0 & 0 & 0 \\ 0 & 0 & 0 \\ 0 & kbe^{-D\tau}P^* & kbe^{-D\tau}S^* \end{pmatrix}$$

The characteristic equation

$$0 = \det(\lambda I - A - e^{-\tau\lambda}B) \tag{8.18}$$

is given by:

$$0 = \begin{vmatrix} \lambda + D + S^* f' & f & 0 \\ -S^* f' & \lambda & kS^* \\ 0 & kP^*(1 - be^{-(D+\lambda)\tau}) & \lambda + D + kS^*(1 - be^{-(D+\lambda)\tau}) \end{vmatrix}$$

where $f = f(R^*)$ and $f' = f'(R^*)$. A tedious calculation reveals that (8.18) has $\lambda + D$ as a factor, the other factor being:

$$\lambda^2 + m\lambda + n + \left(1 - be^{-(D+\lambda)\tau}\right)(p\lambda + q) = 0 \tag{8.19}$$

where

$$m = D + f'(R^*)S^*, \quad n = (kP^* + D)f'(R^*)S^* \tag{8.20}$$
$$p = kS^*, \quad q = kS^*(f'(R^*)S^* - kP^*)$$

For a careful analysis of the characteristic equation (8.18) we refer the reader to [7]. Here we try to determine stability for τ near zero and for τ near τ_c. We begin with the former.

When $\tau = 0$, (8.19) reduces to the pure quadratic

$$\lambda^2 + S^* f'(R^*)\lambda + kS^* P^* \left(f'(R^*) + (b-1)k\right)$$

with positive coefficients. Hence, both roots are negative and it follows that all eigenvalues of $A(0) + B(0)$ are negative. Matrices $A(\tau)$ and $B(\tau)$ are continuous in τ. Therefore, by Theorem 6.8 and Remark 4.5, there exists $\tau_0 \leq \tau_c$ such that if $0 \leq \tau < \tau_0$ all roots of (8.18) have negative real part and E_C is asymptotically stable.

Now we consider the stability of E_C near its bifurcation with E_S for $\tau \approx \tau_c$. As τ decreases through τ_c, one real simple characteristic root $\lambda_S(\tau)$ of the characteristic equation (8.12) for E_S goes from negative to positive and all other roots have negative real part. The characteristic equation (8.18) agrees with (8.12) at $\tau = \tau_c$ because $E_C = E_S$. We seek to determine what happens to the simple real root $\lambda_C(\tau)$ of (8.18) that corresponds to $\lambda_S(\tau)$ for E_S; both vanish at $\tau = \tau_c$. The sign of $\lambda_C(\tau)$ determines the stability properties of E_C because all other roots are near corresponding roots for E_S and hence have negative real parts. The implicit function theorem should lead to determining the sign of this root; it suffices to consider (8.19) inasmuch as the root of interest is determined by it. We view (8.19) as

$$F(\lambda, \tau) = 0$$

where our interest is in the solution $(\lambda_C(\tau)$ satisfying $\lambda_C(0) = 0$ and how it changes with τ. Implicit differentiation gives

$$0 = F_\lambda(0, \tau_c)\frac{d\lambda_C}{d\tau}(\tau_c) + F_\tau(0, \tau_c) \tag{8.21}$$

From (8.19), we find that the critical derivative

$$F_\lambda(0, \tau_c) = m + p(1 - be^{-D\tau_c}) + qb\tau_c e^{-D\tau_c} \tag{8.22}$$
$$= f'(\bar{R})\bar{S}(1 + bk\tau_c e^{-D\tau_c}) > 0$$

is nondegenerate. Similar calculations lead to

$$F_\tau(0, \tau_c) = n_\tau + q_\tau(1 - be^{-D\tau_c}) + qbDe^{-D\tau_c} \tag{8.23}$$
$$= n_\tau - q_\tau \frac{D}{k\bar{S}} + qD\frac{D + k\bar{S}}{k\bar{S}}$$

where subscript τ denotes the partial derivative evaluated at τ_c. From (8.20) and (8.17) we find, after lengthy calculations, that

$$n_\tau = bkD\bar{S}^2\left(f' - \frac{\bar{S}(f')^2 + Df''\bar{S}}{D + \bar{S}f'}\right)e^{-D\tau_c}$$
$$q_\tau = bk^2\bar{S}^3\left(2f' - \frac{D(f''\bar{S} - f')}{D + \bar{S}f'}\right)e^{-D\tau_c}$$

where the argument $\bar{R}$ of f' and f'' is omitted. Inserting these into (8.23), we find that

$$F_\tau(0, \tau_c) = -Dbk\bar{S}^2 f'(\bar{R})e^{-D\tau_c} \tag{8.24}$$

and hence by (8.21), we have

$$\frac{d\lambda_C}{d\tau}(\tau_c) = \frac{Dbk\bar{S}e^{-D\tau_c}}{1 + bk\bar{S}\tau_c e^{-D\tau_c}} > 0 \tag{8.25}$$

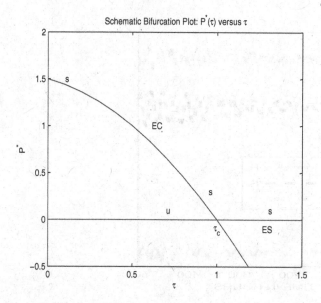

Fig. 8.1 A schematic view of the transcritical bifurcation of E_C from E_S at $\tau = \tau_c$.

Therefore, $\lambda_C(\tau) < 0$ for $\tau < \tau_c$ sufficiently near τ_c. We summarize the results of both calculations in the following result and in Figure 8.1.

Proposition 8.7 *Assume that $bk\bar{S}/(D + k\bar{S}) > 1$ and (8.10) hold. Then there exists τ_i, $i = 0, 1$ satisfying $0 < \tau_0 \leq \tau_1 < \tau_c$ such that E_C is asymptotically stable for $0 \leq \tau < \tau_0$ and for $\tau_1 < \tau < \tau_c$.*

E_C may be unstable as Figure 8.2 shows. Parameters used in the simulation are: $a = 0.0727$, $\tau = 0.5$, $m = .7726$, $b = 25$, $D = 0.2$, $k = 0.024$, and $R_0 = 0.5$. Our simulations incorporated a yield constant of 2×10^{-6}. Initial conditions were chosen near E_C, found by setting $\tau = 0$ and integrating. Undamped oscillations suggest that E_C is unstable and that there is a periodic orbit. See [7] where a Hopf bifurcation analysis is carried out.

We thank Mr. Zhun Han for the bifurcation plot shown in Figure 8.3. His parameter values are: $k = 0.024$, $a = 0.0727$, $m = 0.7726$, $D = 0.2$, and $b = 52$. No yield constant was used. The latent period delay τ appears on the horizontal axis and the maximum and minimum value of the phage density corresponding to a periodic solution or equilibrium appear on the vertical axis.

8.6 Another Formulation of the Model

The form of the initial data (8.5) do not correspond well with experiments described in [50]. In these experiments, phage ($P(0)$) and susceptible bacteria ($S(0)$) are sim-

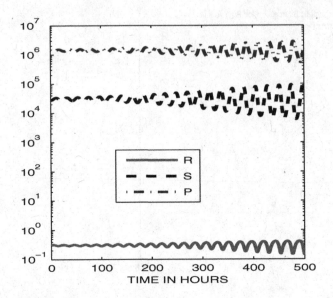

Fig. 8.2 Unstable E_C and periodic solution.

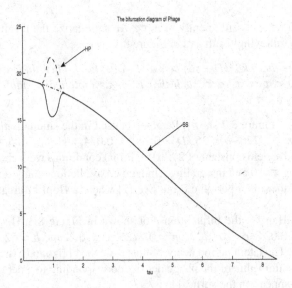

Fig. 8.3 A Hopf bifurcation plot.

ply added to a chemostat at time $t = 0$. One experiment consisted of adding phage to the chemostat after equilibrium E_S had been attained. It seems that no infected cells were included in the initial data, at least insofar as the experimenters were aware. One can try to imagine how to give past histories as in (8.5) which would correspond to these experiments. As $I(0) = 0$ in experiments, (8.3) and the assumed continuity of S and P result in $S(\theta)P(\theta) = 0$, $-\tau \le \theta \le 0$. Obviously, this does not work because this leads to no phage by Proposition 8.6. An alternative is to drop the continuity requirement of past histories.

A more natural formulation of the model (8.4) and initial data that better capture the setup of these and other experiments is described here. It begins with the notion that the appropriate initial data for the problem are to prescribe values for $R(0), S(0), P(0)$, and an infection-age-density for the infected cells introduced at $t = 0$ given by $\iota : [0, \tau] \to \mathbb{R}_+$, where $I(0) = \int_0^\tau \iota(s)ds$ and where

$\iota(s)ds = $ number of infected cells infected between times $-s - ds$ and $-s$

This cohort of infected cells will lyse at time $t = -s + \tau > 0$. Thus s denotes the infection-age of the cohort. In most experiments $\iota = 0$ because no infected cells are added.

Because the infected cells at $t = 0$ are all lysed by $t = \tau$, values of R, S, and P can be found on the first latent interval $0 \le t < \tau$ by solving:

$$R'(t) = D(R_0 - R(t)) - f(R(t))S(t)$$
$$S'(t) = (f(R(t)) - D)S(t) - kS(t)P(t) \qquad (8.26)$$
$$P'(t) = -DP(t) - kS(t)P(t) + be^{-Dt}\iota(\tau - t)$$

with the prescribed values of $R(0), S(0), P(0)$. The term $be^{-Dt}\iota(\tau - t)$ corresponds to the initial infected cohort of infection-age $\tau - t$ which lyse at time t producing b phage per cell. Of course, only the fraction e^{-Dt} of this cohort remains in the chemostat at time t. Note that any new infections produced by the initial phage $P(0)$ will not have matured to lyse during this first latent period. System (8.26) is a system of ordinary differential equations with time-dependent input given by $be^{-Dt}\iota(\tau - t)$ in the equation for P.

For $t \ge \tau$, the initial infected cells, described by ι and numbering $I(0)$, have all lysed or washed out of the chemostat. Hence, values of R, S, P are given by solving the system of delay equations (8.4) using as initial data at $t = \tau$ the data produced by solving (8.26) on the interval $[0, \tau]$. In order to be explicit, we temporarily denote the result of solving (8.26) on the interval $[0, \tau]$ by (R_1, S_1, P_1). Then initial data for (8.4) at $t = \tau$ are given by:

$$R(\tau) = R_1(\tau) \qquad (8.27)$$
$$S(s) = S_1(s), s \in [0, \tau]$$
$$P(s) = P_1(s), s \in [0, \tau]$$

In summary, solutions of (8.26)-(8.4) under our new reformulation are completely determined by the initial data:

$$R(0) \geq 0, S(0) \geq 0, \iota : [0, \tau] \to \mathbb{R}_+, P(0) \geq 0. \qquad (8.28)$$

It suffices that $\iota(\bullet)$ be integrable. Indeed, the equation for P of (8.26) can be integrated using the variation of constants formula to obtain

$$P(t) = P(0)e^{-Dt} + \int_0^t e^{-D(t-\eta)} \left(be^{-D\eta} \iota(\tau - \eta) - kS(\eta)P(\eta) \right) d\eta \qquad (8.29)$$

for $0 \leq t < \tau$. During the first latent period, the number of phage at time t consists of those that were present at time 0 and have neither washed out nor been absorbed prior to time t plus the surviving phage released from the original cohort of infected cells as they lysed.

An important special case of (8.28), which comports with the experiments, is to start at $t = 0$ with no infected cells. If

$$R(0) \geq 0, S(0) > 0, \iota \equiv 0, P(0) > 0$$

then $R(t), S(t), P(t)$ are determined by

$$\begin{aligned}
R'(t) &= D(R_0 - R(t)) - f(R(t))S(t) \\
S'(t) &= (f(R(t)) - D)S(t) - kS(t)P(t) \qquad (8.30) \\
P'(t) &= -DP(t) - kS(t)P(t) + H(t - \tau)be^{-Dt}kS(t - \tau)P(t - \tau)
\end{aligned}$$

where $H(t)$ is the Heaviside function: $H(t) = 0$, $t < 0$, $H(t) = 1$, $t \geq 0$. Notice that inasmuch as there is no source of new phage on the interval $[0, \tau]$, $P(t)$ decreases during this interval as the initial phage are washed out or adhere to susceptible cells.

Figure 8.4 is a simulation of (8.30) initiated by adding 10^6 phage to the equilibrium E_S. Variables have been scaled for better viewing. Parameter values are: $b = 50$, $\tau = 0.5$, $R_0 = 0.5$, $k = 3 \times 10^{-7}$, $D = 0.2$, $m = .7726$, and $a = 0.0727$.

This is all satisfactory from a biological point of view but not from a mathematical viewpoint. What is the state of our dynamical system at time t? Our new formulation of the model suggests that it should be $(R(t), S(t), P(t))$ together with the infection-age distribution of infected cells. We can easily determine

$$i(t, s) = \text{infection-age of infected cells at time } t$$

For $0 \leq t < \tau$, there are two sources of infected cells, namely old cells ($t < s \leq \tau$) which are the survivors of those infected cells introduced at $t = 0$ and young cells ($0 < s \leq t$) which were infected in the interval $(0, t]$ by phage infection. Survivors of the old cells number $e^{-Dt}\iota(s - t)$ because the probability of avoiding being washed out of the chemostat is e^{-Dt} and because these cells have infection-age s at time t. The young cells of infection-age $s \in (0, t]$ were infected at time $t - s$, numbering $e^{-Ds}kS(t - s)P(t - s)$ because they need only survive washout for time s. Hence,

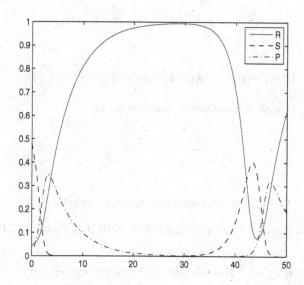

Fig. 8.4 10^6 phage added to E_S at $t = 0$.

during the initial latent period, (i.e., on $0 \le t < \tau$),

$$i(t,s) = \begin{pmatrix} ke^{-Ds}S(t-s)P(t-s), \ 0 \le s < t \\ \iota(s-t)e^{-Dt}, \qquad t < s \le \tau \end{pmatrix} \qquad (8.31)$$

Following the initial latent period, the original cohort has lysed so all infected cells result from infections because $t = 0$. Hence, on $\tau < t$ we have:

$$i(t,s) = ke^{-Ds}S(t-s)P(t-s), 0 \le s \le \tau \qquad (8.32)$$

The function $i : [0,\infty) \times [0,\tau] \to \mathbb{R}_+$ is the solution of the initial-boundary value problem for the partial differential equation

$$\frac{\partial i}{\partial t} + \frac{\partial i}{\partial a} = -Di$$
$$i(t,0) = kS(t)P(t), t > 0 \qquad (8.33)$$
$$i(0,a) = \iota(a), 0 \le a \le \tau$$

See Exercise 8.9. The monograph [26] is recommended for further study.

We must explicitly identify our state space as a metric space and specify the semiflow. Let

$$\tilde{X} = \mathbb{R}_+ \times \mathbb{R}_+ \times L^1_+([0,\tau]) \times \mathbb{R}_+$$

where $L^1_+([0,\tau])$ consists of equivalence classes of nonnegative functions in the space of integrable functions defined on $[0,\tau]$ with norm

$$\|u\|_1 = \int_0^\tau |u(s)|ds$$

Then for $(R(0), S(0), \imath, P(0)) \in \tilde{X}$

$$\check{\Phi}(t, (R(0), S(0), \imath, P(0))) = (R(t), S(t), i(t, \bullet), P(t)), t \geq 0$$

gives the dynamics. It can be shown that Φ is a semiflow on $\tilde{X}$.

Exercises

Exercise 8.1. Verify that (8.3) satisfies the differential equations for I in (8.1).

Exercise 8.2. Show that the virus-free equilibrium (8.9) exists if and only if (8.10) holds.

Exercise 8.3. Show that if $f(R_0) < D$ holds, then $S(t) \to 0$ and $P(t) \to 0$ for (8.4). Use $\limsup_{t \to \infty} R(t) \leq R_0$.

Exercise 8.4. Verify that all hypotheses of Theorem 5.17 hold for V, given in the proof of Proposition 8.6.

Exercise 8.5. Verify that (8.14) holds in case f is of Monod type $f(R) = mR/(a + R)$.

Exercise 8.6. Verify (8.17).

Exercise 8.7. If phage attachment to infected cells is included in the model, the last two equations become:

$$I'(t) = kS(t)P(t) - DI(t) - e^{-D\tau}kS(t - \tau)P(t - \tau)$$
$$P'(t) = -DP(t) - k(S(t) + I(t))P(t) + be^{-D\tau}kS(t - \tau)P(t - \tau)$$

Note that there is no term $-kPI$ in the I equation because phage attachment to an infected cell does not change that cell's status; it results only in a loss of phage. Show that the sign of $PRN - 1$ still determines local stability of the phage-free equilibrium. Show that Proposition 8.3 still holds.

Exercise 8.8. A discrete delay might be replaced by a distributed delay in (8.4). Show that the system of ODEs:

$$R' = D(R_0 - R) - f(R)S$$
$$S' = (f(R) - D)S - kSP$$
$$I'_1 = akSP - (a + D)I_1$$
$$I_j = aI_{j-1} - (a + D)I_j, 2 \leq j \leq p$$
$$P' = -kPS - DP + bI_p$$

leads to the distributed delay system

$$R' = D(R_0 - R) - f(R)S$$
$$S' = (f(R) - D)S - kSP$$
$$P' = -kPS - DP + bk \int_0^\infty P(t-s)S(t-s)e^{-Ds}g_a^p(s)ds$$

Hint: Set

$$I_j(t) = \int_0^\infty kP(t-s)S(t-s)e^{-Ds}g_a^j(s)ds$$

and note that $e^{-Ds}g_a^j(s) \propto g_{a+D}^j(s)$.

Exercise 8.9. Show that the infection-age distribution $i(t,s)$ satisfies (8.33), except possibly along the line $t = s$. Give conditions for it to be continuous along this line and to satisfy (8.33) in $(0,\infty) \times (0,\tau]$.

Exercise 8.10. Suppose a mutant bacteria arises that is resistant to phage infection. Write down the system of differential equations that include the mutant and the susceptible strain, both consuming nutrient, but assume that the mutant is less fit than the susceptible strain in having a reduced growth rate. Is there an equilibrium in which phage, mutant, and susceptible strains coexist? See [50, 51].

Appendix A
Results from Real and Complex Analysis

Abstract Some key results from real and complex analysis are reviewed here. These include a brief introduction to analytic functions of a complex variable with special emphasis on Rouché's theorem and the implicit function theorem. The Ascoli–Arzela lemma has special significance here as our dynamics takes place in the space of continuous functions on an interval. The fluctuation lemma and the Gronwall lemma are stated.

A.1 Analytic Functions

A good elementary reference for the material on complex functions is [17]. We also reference results in [1].

The complex plane $\mathbb{C}$ is the set of all complex numbers $z = x + iy$. The real and imaginary parts of z are defined by $\Re(z) = x$ and $\Im(z) = y$. The modulus of z is $|z| = (x^2 + y^2)^{1/2}$. We often identify $\mathbb{C}$ with the (x, y)-plane, the complex number $z = x + iy$ being identified with the point (x, y). The complex conjugate of z is its reflection in the x-axis: $\bar{z} = x - iy$. Conjugation has nice properties; the conjugate of the sum, product and quotient of two complex numbers is the sum, product or quotient, respectively, of the conjugates.

A complex-valued function of a complex variable $f : D \to \mathbb{C}$, where $D \subset \mathbb{C}$, can be represented as

$$w = f(z) = u(z) + iv(z) = u(x, y) + iv(x, y)$$

where u and v are real-valued functions defined on the domain D, now viewed as a subset of the (x, y)-plane. An example is the exponential function

$$w = e^z = e^x \cos(y) + ie^x \sin(y)$$

Here, $u = e^x \cos(y)$ and $v = e^x \sin(y)$. The reader should verify that $|e^z| = e^x = e^{\Re(z)}$ and that $\overline{e^z} = e^{\bar{z}}$. Thus, $e^z \neq 0$.

We say that f is analytic on D provided D is an open set and f is differentiable at each point of D in the sense that

$$f'(z_0) = \lim_{z \to z_0} \frac{f(z) - f(z_0)}{z - z_0}$$

exists at each $z_0 \in D$. If f is analytic on all of $\mathbb{C}$ then f is said to be an entire function. The derivative has the usual properties we learn for functions of a real variable; rules for differentiating are similar and the usual formulas hold for derivatives of polynomial functions $f(z) = a_0 z^n + a_1 z^{n-1} + \cdots + a_n$, and the standard functions from calculus $e^z, \cos(z), \sin(z)$.

One can verify analyticity of f directly from its real and imaginary parts u and v.

Theorem A.1 *If f is analytic in D then u, v have partial derivatives u_x, u_y and v_x, v_y at all points of D that satisfy the Cauchy–Riemann equations*

$$u_x = v_y, \ u_y = -v_x \tag{A.1}$$

Moreover,

$$f'(z) = u_x + iv_x = v_y - iu_y$$

Conversely, if u_x, u_y and v_x, v_y exist in D and are continuous and satisfy (A.1), *then f is analytic in D.*

Exercise A.1. Verify that the complex exponential function is analytic in $\mathbb{C}$.

The Cauchy integral theorem [17] is the truly remarkable fact about analytic functions. We do not state it here but we make use of some of its consequences:

(a) An analytic function is infinitely differentiable.
(b) The Taylor series expansion of an analytic function converges and represents that function.

If $f : D \to \mathbb{C}$ is analytic, then all its derivatives:

$$f'(z), f''(z), \ldots, f^{(n)}(z), \ldots$$

exist at every point of D. Moreover, the Taylor series centered at $z_0 \in D$:

$$f(z) = \sum_{n=0}^{\infty} \frac{f^{(n)}(z_0)}{n!} (z - z_0)^n \tag{A.2}$$

converges to $f(z)$ for all z satisfying $|z - z_0| < R$ as long as $\{z : |z - z_0| < R\} \subset D$. In fact, the series converges absolutely (i.e., the series obtained by taking term-by-term modulus converges).

Conversely, if a power series

$$\sum_{n=0}^{\infty} a_n (z - z_0)^n$$

converges for some $z = z_1$, then it converges absolutely for all z satisfying $|z - z_0| < |z_1|$ to an analytic function.

One notable consequence of these facts is that an analytic function defined on a connected open domain D in $\mathbb{C}$ and vanishing at the point $z_0 \in D$ is either the identically zero function or it has a zero of finite order at z_0. Recall, we say that z_0 is a zero of order $k \geq 1$ of f if

$$f(z_0) = f'(z_0) = \cdots f^{(k-1)}(z_0) = 0, f^{(k)}(z_0) \neq 0$$

If f has a zero of order k at z_0, then from the power series we see that

$$f(z) = \sum_{n=k}^{\infty} \frac{f^{(n)}(z_0)}{n!}(z - z_0)^n = (z - z_0)^k g(z)$$

where

$$g(z) = \sum_{n=0}^{\infty} \frac{f^{(n+k)}(z_0)}{(n+k)!}(z - z_0)^n$$

is an analytic function. Moreover, $g(z_0) = f^{(k)}(z_0) \neq 0$ so by continuity of g there is a neighborhood U of z_0 such that $g(z) \neq 0$, $z \in U$. It follows that $f(z) \neq 0$ for $z \in U \setminus \{z_0\}$.

Summarizing, an analytic function that is not identically zero in its (connected) domain has isolated zeros. This fact has important consequences.

Proposition A.2 *Let f be analytic on a connected domain D, not identically zero in D, and let K be a closed and bounded subset of D. Then f has at most finitely many zeros in K. If f is an entire function, then it has at most countably many zeros; if it has infinitely many zeros and $\{z_n\}_{n=1}^{\infty}$ is an enumeration of its distinct zeros, then $|z_n| \to \infty$ as $n \to \infty$.*

Proof. If there were infinitely many zeros in K we could find a sequence $\{z_n\} \subset K$ of distinct points such that $f(z_n) = 0$. By Bolzano–Weierstrass theorem, this sequence must have a limit point. Thus, there must be a convergent subsequence, which we rename as $\{z_n\}$, such that $z_n \to z \in K$. By continuity of f it follows that $f(z) = 0$ and thus f has a nonisolated zero. Because f is not identically zero, this gives a contradiction.

If f is entire, then it has finitely many zeros in each closed ball $\{z : |z| \leq n\}$ so it has at most countably many zeros and at most finitely many of these may lie inside any closed ball $\{z : |z| \leq R\}$ where $R > 0$. $\square$

A.2 Implicit Function Theorem for Complex Variables

One of our main tasks in determining the stability of an equilibrium solution is to understand the characteristic roots of an analytic characteristic equation $h(z) = 0$. In practice, there are usually important parameters, such as the delay, and we would like to know how the roots vary with the parameters. Therefore, we must study the roots z of the equation

$$h(z,p) = 0 \qquad\qquad (A.3)$$

where p denotes a vector of usually real parameters. The implicit function theorem is the natural tool for this. The following is an adaptation of the usual implicit function theorem (Theorem 9.28, [65]) to complex-valued functions.

Theorem A.3 *Let $h : D \times O \to \mathbb{C}$ where $D \subset \mathbb{C}$ and $O \subset \mathbb{R}^k$ are both open sets. Assume that h is analytic in $z \in D$ for each $p \in O$ and $h_z(z,p)$ is continuous in $D \times O$. Assume also that $h_p(z,p)$ exists and is continuous in $D \times O$.*

If $h(z_0,p_0) = 0$ for some $(z_0,p_0) \in D \times O$ and $h_z(z_0,p_0) \neq 0$, then there is a neighborhood U of z_0 in D and a neighborhood V of p_0 in O and a continuously differentiable function $g : V \to U$ satisfying:

(a) $g(p_0) = z_0$.

(b) $h(g(p),p) = 0$, $p \in V$.

(c) If $(z,p) \in U \times V$ and $h(z,p) = 0$, then $z = g(p)$.

Proof. To see that this follows from the usual implicit function theorem (Theorem 9.28, [65] or see Theorem A.5), we identify (A.3) with

$$H(x,y,p) = 0, H(x,y,p) = (u(x,y,p), v(x,y,p))$$

where $h(z,p) = u(x,y,p) + iv(x,y,p)$ and $z = x + iy$. H is continuously differentiable on its domain. Then we have $H(x_0,y_0,p_0) = 0$ and its Jacobian with respect to (x,y) satisfies

$$\frac{\partial H}{\partial (x,y)}(x_0,y_0,p_0) = \begin{pmatrix} u_x & u_y \\ v_x & v_y \end{pmatrix} = \begin{pmatrix} u_x & -v_x \\ v_x & u_x \end{pmatrix}$$

where the partial derivatives are evaluated at (x_0,y_0,p_0) and where we have used the Cauchy–Riemann equations (A.1). The determinant of the Jacobian is given by $u_x^2 + v_x^2 = |h_z(z_0,p_0)|^2$ because $h_z = u_x + iv_x$ by Theorem A.1. By hypothesis, $h_z(z_0,p_0) \neq 0$ so the determinant is nonzero as required. $\square$

A.3 Rouché's Theorem

The following result, a special case of Rouché's theorem (see ([1]), is useful in studying the characteristic equation.

Theorem A.4 *[Rouché's theorem] Let γ be a simple closed curve (non-intersecting) in the complex plane and let $f(z)$ and $g(z)$ be functions analytic in the complex plane and satisfying*

$$|f(z) - g(z)| < |f(z)|, z \in \gamma$$

Then $f(z)$ and $g(z)$ have the same number of zeros, counting the order of each root, enclosed by γ.

To see why Rouché's theorem is useful, keep in mind that in practice our linear systems always contain many parameters so usually we have $F(z) = F(p,z)$ where $p \in \mathbb{R}^k$ are parameters and we want to know how the characteristic zeros vary as p is varied. Let's suppose that F is continuous in all arguments but analytic in z for fixed p. Suppose that γ is a simple closed curve in $\mathbb{C}$ and $F(z,p_0) = 0$ has no roots on γ for parameter value p_0. Now compactness of γ and continuity of F mean that

$$\varepsilon := \min\{|F(z,p_0)| : z \in \gamma\} > 0$$

For the same reasons, there exists $\delta > 0$ such that

$$|p - p_0| < \delta, z \in \gamma \Longrightarrow |F(p,z) - F(p_0,z)| < \varepsilon$$

from which we conclude, by Theorem A.4, that the number of roots of $F(p,z) = 0$ inside γ is the same as the number of roots of $F(p_0,z) = 0$ provided $|p - p_0| < \delta$.

Exercise A.2. Use Theorem A.4 to prove that if $p(z)$ is a polynomial of degree n and $\varepsilon > 0$ is such that $p(z) = 0$ has a root z_0 of multiplicity m and no other roots in $|z - z_0| \leq \varepsilon$, then there exists $\delta > 0$ such that nth degree polynomial $q(z)$ has m zeros, counting multiplicity, in $|z - z_0| \leq \varepsilon$ provided the coefficients of q are within δ of those of p.

A.4 Ascoli–Arzela Theorem

Let $C = C([-r,0], \mathbb{R})$ be the metric space of continuous real-valued functions on the interval $[-r,0]$ with the supremum norm

$$\|\phi\| = \sup\{|\phi(x)| : -r \leq x \leq 0\}$$

We use the argument x rather than θ for ϕ for simplicity. A sequence $\{\phi_n\}_{n=1}^{\infty}$ in C converges to $\phi \in C$ relative to the supremum norm if and only if it converges uniformly on $[-r,0]$: $\forall \varepsilon > 0, \exists$ a natural number N such that

$$|\phi_n(x) - \phi(x)| < \varepsilon, x \in [-r,0], n \geq N$$

It is important to know when a given sequence $\{\phi_n\}_{n=1}^{\infty}$ in C has a convergent subsequence. The Bolzano–Weierstrass theorem for $\mathbb{R}^n$ says that every bounded sequence of vectors has a convergent subsequence. However, this property fails for continuous function spaces. For example, consider the sequence $\phi_n(x) = x^n$ in $C([0,1], \mathbb{R})$ for $n = 1,2,\cdots$. As $|\phi_n(x)| \leq 1$, $x \in [0,1]$, $n \geq 1$, it is a bounded sequence but it has no subsequence that converges uniformly on $[0,1]$ to a member of $C([0,1], \mathbb{R})$. Indeed, because $\phi_n(x)$ converges pointwise to 0 if $x < 1$ and to 1 if

$x = 1$, so will any subsequence. Therefore, in the space C, we need additional conditions beside boundedness to guarantee the existence of a uniformly convergent subsequence.

A subset A of functions in C is *equicontinuous* if for every $\varepsilon > 0$ there exists $\delta > 0$ such that $|\phi(x) - \phi(y)| < \varepsilon$ whenever $\phi \in A$ and $x, y \in [-r, 0]$ satisfy $|x - y| < \delta$. Note that the same δ works for every $\phi \in A$ and every $x, y \in [-r, 0]$ with $|x - y| < \delta$. The most common method of verifying equicontinuity is to show that there exists $M > 0$ such that ϕ' exists and $|\phi'(x)| \leq M$ for every $\phi \in A$ and every $x \in [-r, 0]$. Then A is equicontinuous because

$$|\phi(x) - \phi(y)| = |\phi'(\eta)||x - y| \leq M|x - y|$$

holds for every $\phi \in A$ by the mean value theorem, where $\eta \in [-r, 0]$ may depend on $\phi \in A$.

We require the famous Ascoli–Arzela theorem, Theorem 7.25 [65].

Theorem A.1. *Let $\{\phi_n\}_{n=1}^{\infty}$ be a sequence of functions in C that is equicontinuous and such that there is some $M > 0$ such that $|\phi_n(x)| \leq M$ for all $n \geq 1$ and all $x \in [-r, 0]$. Then some subsequence of $\{\phi_n\}_{n=1}^{\infty}$ converges uniformly to an element of C.*

We remark that by replacing absolute value by a vector norm on $\mathbb{R}^n$, the definitions above and Theorem A.1 extend to $C([-r, 0], \mathbb{R}^n)$.

A.5 Fluctuation Lemma

Let $f : [b, \infty) \to \mathbb{R}$. Then the *limit superior* and the *limit inferior* of f as $t \to \infty$ are defined as

$$f^{\infty} := \limsup_{t \to \infty} f(t) = \inf_{r \geq b} \sup\{f(t); t \geq r\}$$

$$f_{\infty} := \liminf_{t \to \infty} f(t) = \sup_{r \geq b} \inf\{f(t); t \geq r\} \tag{A.4}$$

It is easily shown that there is a sequence $s_k \to \infty$ such that $f(s_k) \to f_{\infty}$ and that there is a sequence $t_k \to \infty$ such that $f(t_k) \to f^{\infty}$. In fact, f_{∞} is the smallest such sequential limit and f^{∞} is the largest.

Perhaps the most useful property of f^{∞} is that for every $\varepsilon > 0$, there exists $T > b$ such that $f(t) \leq f^{\infty} + \varepsilon$ for all $t \geq T$. Analogously, for every $\varepsilon > 0$, there exists $T > b$ such that $f(t) \geq f_{\infty} - \varepsilon$ for all $t \geq T$.

The following result, often called the fluctuation lemma, is remarkably useful. It is intuitive if one thinks of $f(t) = \sin t$ where $f_{\infty} = -1$ and $f^{\infty} = 1$. See [37, 75, 72] for a proof or the reader can supply it.

Lemma A.1. *Let $f : [b, \infty) \to \mathbb{R}$ be bounded and differentiable. Then there exist sequences $s_k, t_k \to \infty$ such that*

$$\left.\begin{array}{ll} f(s_k) \to f_\infty, & f'(s_k) \to 0 \\ f(t_k) \to f^\infty, & f'(t_k) \to 0 \end{array}\right\} k \to \infty.$$

A.6 General Implicit Function Theorem

In the following appendix, we require the implicit function theorem in a Banach space setting. We follow [81]; see also [16]. Recall that a Banach space X is a complete normed linear space; we use the notation $\| \bullet \|_X$ for the norm on X. A mapping F is said to be C^m, written $F \in C^m$, if it is m-times continuously differentiable.

Theorem A.5 *Suppose that mapping $F : U(x_0, y_0) \subset X \times Y \to Z$ is defined on an open set $U(x_0, y_0)$ and $F(x_0, y_0) = 0$, where X, Y, and Z are Banach spaces. Assume that the partial derivative $F_y(x, y)$ exists for $(x, y) \in U(x_0, y_0)$, F and F_y are continuous at (x_0, y_0), and that $F_y(x_0, y_0) : Y \to Z$ is bijective. Then:*

(a) There exist $r_0, r > 0$ such that for every $x \in X$ with $\|x - x_0\|_X \le r_0$, there is exactly one $y = y(x) \in Y$ for which $\|y - y_0\|_Y \le r$ and $F(x, y(x)) = 0$.
(b) If F is continuous in $U(x_0, y_0)$, then $y(x)$ is continuous in a neighborhood of x_0.
(c) If $F \in C^m$, $1 \le m \le \infty$ on $U(x_0, y_0)$, then $y(\bullet) \in C^m$ on some neighborhood of x_0.

The standard implicit function theorem from advanced calculus is the special case $X = \mathbb{R}^m, Y = Z = \mathbb{R}^n$. The proof is the same for both the finite and infinite-dimensional cases.

A.7 Gronwall's Inequality

We recall a fundamental result from ODE theory that plays more or less the same role for delay differential equations. See [40, 10] for the elementary proof.

Theorem A.6 (Gronwall Inequality) *Let $K \ge 0$ and let $f, g : [a, b] \to [0, \infty)$ be continuous functions satisfying the inequality*

$$f(t) \le K + \int_a^t f(s)g(s)ds, a \le t \le b.$$

Then

$$f(t) \le K \exp(\int_a^t g(s)ds), a \le t \le b.$$

Appendix B
Hopf Bifurcation for Delayed Negative Feedback

Abstract In Chapter 6.3, we gave a purely formal construction of the Hopf bifurcation of periodic solutions of the canonical nonlinear negative feedback equation $x'(t) = -f(x(t-r))$. Here we give a rigorous proof of the existence of these solutions using the implicit function theorem A.5.

B.1 Basic Setup and Preliminaries

In this appendix, we give a mathematically rigorous justification of the formal arguments given in Chapter 6 for the computation of the Hopf periodic solution of the scalar delay equation with negative feedback. We begin with Equation (6.13).

We seek solutions (P, R, ω) of

$$P'(z) = -\frac{R}{\omega} f(P(z - \frac{\pi}{2}\omega)) \tag{B.1}$$

where P is 2π-periodic and

$$P \approx 0, R \approx 1, \omega \approx 1$$

Our immediate goal is to reformulate this differential equation as an equation in a suitable Banach space. For this, we need some notation and preliminary work. It is better to rewrite the equation as

$$L(P)(z) = P(z - \frac{\pi}{2}) - \frac{R}{\omega} f(P(z - \frac{\pi}{2}\omega)) \tag{B.2}$$

where

$$LP(z) = P'(z) + P(z - \frac{\pi}{2})$$

P' denotes the derivative of P. We begin by studying this linear operator.

Fourier series play a big role.

$$h = \sum_{n \in Z} h_n e^{inz}$$

is the Fourier series for h, where

$$h_n = \frac{1}{2\pi} \int_{-\pi}^{\pi} h(z) e^{-inz} dz, n \in \mathbb{Z}$$

It converges to h in the mean square sense. We use complex series mainly for the linear equation. When we consider the nonlinear equation, we use real Fourier series.

For $k \geq 0$, let

$$H^k = \{h \in L^2(\mathbb{T}) : \sum_{n \in Z} n^{2k} |h_n|^2 < \infty\}$$

These spaces are Hilbert spaces contained in the Hilbert space $H^0 = L^2(\mathbb{T})$ of square integrable functions on the unit circle $\mathbb{T}$. Let

$$C^k = \{f : \mathbb{R} \to \mathbb{R} : f \text{ is k times continuously differentiable and } 2\pi - \text{periodic}\}$$

denote the Banach space with norm

$$\|f\|_k = \|f\|_\infty + \|f'\|_\infty + \cdots + \|f^{(k)}\|_\infty$$

where $\| \cdot \|_\infty$ denotes the supremum norm. It is well known that

$$H^{k+1} \subset C^k \subset H^k, \ k = 0, 1, 2, \ldots$$

Proposition B.1 *Let $k \geq 1$. Then $L : H^{k+1} \to H^k$ is a bounded linear operator and*

$$N(L) = span\{\cos(z), \sin(z)\}, \ R(L) = M^k := \{Q \in H^k : Q_1 = Q_{-1} = 0\}$$

where $N(L)$ denotes null space and $R(L)$ the image of L. $L^{-1} : M^k \to M^{k+1}$ is bounded.

$L : C^{k+1} \to C^k$ *is a bounded linear operator and*

$$N(L) = span\{\cos(z), \sin(z)\}, \ R(L) = Z^k := \{Q \in C^k : Q_1 = Q_{-1} = 0\}$$

$L^{-1} : Z^k \to Z^{k+1}$ *is bounded.*

Proof. Consider

$$P'(z) + P(z - \frac{\pi}{2}) = h(z) \in H^k$$

Both h and the solution P have Fourier series and we can solve for P if we can determine its Fourier coefficients in terms of those of h. The relevant series are:

$$h = \sum_{n \in Z} h_n e^{inz}, \; P = \sum_{n \in Z} P_n e^{inz}$$

$$P(\bullet - \frac{\pi}{2}) = \sum_{n \in Z} P_n e^{-in\frac{\pi}{2}} e^{inz}$$

$$\frac{dP}{dz} = \sum_{n \in Z} in P_n e^{inz}$$

Inserting these into our equation and equating coefficients of e^{inz} leads to

$$(LP)_n = (in + e^{-in\pi/2})P_n = h_n, \; n \in Z$$

Consequently, as the term in parentheses vanishes if and only if $n = \pm 1$, we find that the null space of L, $N(L)$ is spanned by e^{iz}, e^{-iz} and that there is a 2π-periodic solution P if and only if

$$h_1 = h_{-1} = 0 \tag{B.3}$$

and its Fourier coefficients are given by

$$P_n = \frac{h_n}{in + (-i)^n}, |n| > 1, P_0 = h_0 \tag{B.4}$$

and where P_1 and P_{-1} are arbitrary. Thus,

$$(L^{-1}h)_n = \frac{h_n}{in + (-i)^n}, n \neq \pm 1$$

It is a straightforward exercise to establish the assertions regarding $L : H^{k+1} \to H^k$. Clearly, $L(C^{k+1}) \subset C^k$. If $h \in C^k$ satisfies $h_1 = h_{-1} = 0$, then $LP = h$ has a solution $P \in H^{k+1}$ because $h \in H^k$. It follows that $P(z - \pi/2)$ belongs to C^k because $H^{k+1} \in C^k$ and as $P'(z) = -P(z - \pi/2) + h(z)$, we conclude that $P' \in C^k$ implying that $P \in C^{k+1}$. $\square$

Define the projection operator $Q : C^k \to C^k$ by

$$QP = P_{-1}e^{-iz} + P_1 e^{iz}.$$

The following result is a reformulation that is more useful for solving equations.

Corollary B.2 *The equation*
$$LP = h \in C^k$$

has a solution $P \in C^{k+1}$ if and only if

$$Qh = 0$$

and

$$LP = (I - Q)h$$

The last equation has a unique solution satisfying $QP = 0$.

We need to know that substitution operators are smooth.

Lemma B.1. *Let $f : \mathbb{R} \to \mathbb{R}$ have continuous derivatives of all order. Then the map $F : C^k \to C^k$ given by $P \to f(P)$ is continuously differentiable and its derivative is given by*

$$DF(P)(h)(t) = f'(P(t))h(t), \ h \in C^k$$

Proof. We merely point out what needs to be shown, namely,

$$\lim_{\|h\|_k \to 0} \frac{\|F(P+h) - F(P) - DF(P)h\|_k}{\|h\|_k} = 0$$

For example, if $k = 2$, one can show that

$$g(t) = f(P(t) + h(t)) - f(P(t)) - f'(P(t))h(t)$$

satisfies

$$\|g\|_\infty + \|g'\|_\infty + \|g''\|_\infty = O((\|h\|_\infty + \|h'\|_\infty + \|h''\|_\infty)^2)$$

by simple, but tedious, calculations. $\quad\square$

For $\theta \in \mathbb{R}$, define the shift operator $T_\theta : C^k \to C^k$ by

$$[T_\theta P](z) = P(z - \theta)$$

Then $\{T_\theta\}_{\theta \in \mathbb{R}}$ defines a group of bounded linear operators on the spaces H^k and C^k. Note that

$$T_\theta L = L T_\theta, T_\theta Q = Q T_\theta.$$

C^k is an algebra and T_θ is an algebra homomorphism:

$$T_\theta(P + Q) = T_\theta P + T_\theta Q, \quad T_\theta(P \cdot Q) = T_\theta P \cdot T_\theta Q$$

More generally, we use that $T_\theta f(P) = f(T_\theta P)$ for $P \in C^k$.

There is a certain loss of smoothness in applying the shift map which is immediately apparent from its definition above: to differentiate with respect to θ we must differentiate P.

Lemma B.2. *The map $K : \mathbb{R} \times C^{k+1} \to C^k$ given by $(\theta, P) \to T_\theta P$ is C^1 and*

$$DK(\theta, P)(h, Q) = T_\theta(P'h + Q)$$

B.2 The Solution

In order to simplify Equation (B.2), we write

$$\frac{\pi}{2}\omega = \frac{\pi}{2} + \delta, \frac{R}{\omega} = 1 + \mu \tag{B.5}$$

where $\delta, \mu \approx 0$. Write

$$f(u) = u + u^2 G(u)$$

where $G(0) = A/2$ and G is smooth.

Then (B.2) becomes

$$LP = T_{\pi/2}P - (1+\mu)T_{(\pi/2+\delta)}\left(P + P^2 G(P)\right)$$

Note that we have dropped arguments of functions (e.g., the z variable) to emphasize that we are now seeking an abstract formulation. We continue to do this although it leads to writing cos instead of $\cos(z)$; cos belongs to C^k but $\cos(z)$ is a scalar belonging to $\mathbb{R}$.

We seek a solution $P = \varepsilon(\cos + q)$ where $q \in Z^2$. As $L\cos = 0$ and $T_{\pi/2}\cos = \sin$, this becomes

$$\varepsilon Lq = \varepsilon \sin + \varepsilon T_{\pi/2}q - (1+\mu)T_{(\pi/2+\delta)}(\varepsilon \cos + \varepsilon q$$
$$+ (\varepsilon \cos + \varepsilon q)^2 G(\varepsilon \cos + \varepsilon q))$$

According to Corollary B.2, and using that $Qq = 0$ and $Q\sin = \sin$ and $Q\cos = \cos$, this equation is equivalent to the following system:

$$0 = \varepsilon \sin - (1+\mu)T_{(\pi/2+\delta)}\left(\varepsilon \cos + \varepsilon^2 Q(\cos + q)^2 G(\varepsilon \cos + \varepsilon q)\right) \qquad (B.6)$$
$$\varepsilon Lq = \varepsilon T_{\pi/2}q - (1+\mu)T_{(\pi/2+\delta)}\left(\varepsilon q + \varepsilon^2(I-Q)(\cos + q)^2 G(\varepsilon \cos + \varepsilon q)\right)$$

B.2.1 Solve for q

We consider the equation for $q \in Z^2$ first. Dividing by ε, it becomes

$$Lq = T_{\pi/2}q - (1+\mu)T_{(\pi/2+\delta)}\{q + \varepsilon(I-Q)(\cos + q)^2 G(\varepsilon \cos + \varepsilon q)\} \qquad (B.7)$$

This equation has an important symmetry: if $(\mu, \delta, \varepsilon, q)$ satisfies this equation, where $q \in Z^2$, then so does $(\mu, \delta, -\varepsilon, -T_\pi q)$. To see this, first apply T_π to both sides, use the fact that it is an algebra homomorphism, that $T_\pi \cos = -\cos$, and that it commutes with Q, then multiply both sides by -1. These steps are carried out below where $\tilde{q} = -T_\pi q$:

$$LT_\pi q = T_{\pi/2}T_\pi q - (1+\mu)T_{(\pi/2+\delta)}\{T_\pi q$$
$$+ \varepsilon(I-Q)(-\cos + T_\pi q)^2 G(-\varepsilon \cos + \varepsilon T_\pi q)\}$$
$$L(-T_\pi q) = T_{\pi/2}(-T_\pi q) - (1+\mu)T_{(\pi/2+\delta)}\{(-T_\pi q)$$
$$+ (-\varepsilon)(I-Q)(-\cos -(-T_\pi q))^2 G((-\varepsilon)\cos + (-\varepsilon)(-T_\pi q))\}$$
$$L\tilde{q} = T_{\pi/2}\tilde{q} - (1+\mu)T_{(\pi/2+\delta)}\{\tilde{q}$$
$$+ (-\varepsilon)(I-Q)(\cos + \tilde{q})^2 G((-\varepsilon)\cos + (-\varepsilon)\tilde{q})\}$$

By virtue of Lemma B.2 and Lemma B.1, we may view the right side of (B.7) as a C^1 map taking $(\mu,\delta,\varepsilon,q) \in \mathbb{R}^3 \times Z^2$ into Z^1. Lemma B.2 accounts for the loss of one derivative. Hence, this equation is equivalent to

$$0 = q - L^{-1}[T_{\pi/2}q - (1+\mu)T_{(\pi/2+\delta)}\{q + \varepsilon(I - Q)(\cos + q)^2 G(\varepsilon \cos + \varepsilon q)\}] \quad \text{(B.8)}$$

We view (B.8) as

$$F(\mu,\delta,\varepsilon,q) = 0$$

where $F : \mathbb{R}^3 \times Z^2 \to Z^2$ is a C^1 map satisfying

$$F(0,0,0,0) = 0, F_q(0,0,0,0) = I$$

where F_q denotes the Frechet derivative with respect to q and I is the identity. This derivative, and other partial derivatives computed hereafter, are best computed by "freezing" the other variables at their designated values first. For example, $F_q(0,0,0,0)$ is computed by first computing $F(0,0,0,q)$, then its derivative with respect to q.

In fact, F satisfies

$$F(\mu,\delta,0,0) = 0 \quad \text{(B.9)}$$

By the implicit function theorem A.5, there exists a C^1 function $q : \mathbb{R}^3 \to Z^2$, defined near $(0,0,0)$ such that $q = q(\mu,\delta,\varepsilon)$ satisfies

$$F(\mu,\delta,\varepsilon,q(\mu,\delta,\varepsilon)) = 0, \text{ and } q(0,0,0) = 0$$

By the symmetry mentioned above and the uniqueness of solutions guaranteed by the implicit function theorem, we must have that

$$T_\pi q(\mu,\delta,\varepsilon) = -q(\mu,\delta,-\varepsilon) \quad \text{(B.10)}$$

and by (B.9)

$$q(\mu,\delta,0) = 0$$

It follows that

$$q_\mu(0,0,0) = q_\delta(0,0,0) = 0$$

A straightforward calculation gives:

$$F_\varepsilon(0,0,0,0) = G(0)L^{-1}(I - Q)\sin^2$$
$$= G(0)[\frac{1}{2} + \frac{1}{10}\cos(2z) - \frac{1}{5}\sin(2z)]$$

Implicit differentiation of $F = 0$ yields:

$$0 = F_\varepsilon(0,0,0,0) + F_q(0,0,0,0)q_\varepsilon = F_\varepsilon(0,0,0,0) + q_\varepsilon$$

which implies that

$$q_\varepsilon = -G(0)[\frac{1}{2} + \frac{1}{10}\cos(2z) - \frac{1}{5}\sin(2z)] \qquad (B.11)$$

B.2.2 Solve for μ and δ

Having solved the second of equations (B.6) for $q = q(\mu, \delta, \varepsilon)$, we now insert this into the first equation and try to solve it for (μ, δ) in terms of ε. As the right-hand side of the second equation belongs to $R(Q) = \text{span}\{\cos, \sin\}$, it actually represents a system of two equations for the sin- and cos-components of the right-hand side. First, we divide out a factor of ε from the equation to get

$$0 = \sin - (1+\mu)T_{(\pi/2+\delta)}\left(\cos + \varepsilon Q(\cos + q)^2 G(\varepsilon\cos + \varepsilon q)\right)$$

where $q = q(\mu, \delta, \varepsilon)$. Now we break this up into components. As $T_{(\pi/2+\delta)}\cos = \cos\delta\sin - \sin\delta\cos$, we obtain the two equations for $(\mu, \delta, \varepsilon)$:

$$0 = 1 - (1+\mu)\cos\delta - \varepsilon(1+\mu)[T_{(\pi/2+\delta)}(\cos + q)^2 G(\varepsilon\cos + \varepsilon q)]_{sin}$$
$$0 = (1+\mu)\sin\delta - \varepsilon(1+\mu)[T_{(\pi/2+\delta)}(\cos + q)^2 G(\varepsilon\cos + \varepsilon q)]_{cos}$$

where

$$[h]_{sin} = \frac{1}{\pi}\int_{-\pi}^{\pi} h(z)\sin(z)dz$$

denotes the sin-component and $[h]_{cos}$ is similar, with cos replacing sin. Note that operator Q would be redundant if left in place, so it has been removed. Finally, we divide out the factor $(1+\mu)$ so we have

$$0 = -\frac{\mu}{1+\mu} + 1 - \cos\delta - \varepsilon h(\mu, \delta, \varepsilon) \qquad (B.12)$$
$$0 = \sin\delta - \varepsilon k(\mu, \delta, \varepsilon)$$

where

$$h(\mu, \delta, \varepsilon) = [T_{(\pi/2+\delta)}(\cos + q)^2 G(\varepsilon\cos + \varepsilon q)]_{sin} \qquad (B.13)$$
$$k(\mu, \delta, \varepsilon) = [T_{(\pi/2+\delta)}(\cos + q)^2 G(\varepsilon\cos + \varepsilon q)]_{cos} \qquad (B.14)$$

Using (B.10) and that $[T_\pi P]_{cos} = -[P]_{cos}$ for $P \in C^k$, we see that k is odd:

$$k(\mu, \delta, -\varepsilon) = [T_{(\pi/2+\delta)}(\cos - T_\pi q)^2 G(-\varepsilon\cos + \varepsilon T_\pi q)]_{cos}$$
$$= [T_{(\pi/2+\delta)}T_\pi(-\cos - q)^2 T_\pi G(\varepsilon\cos + \varepsilon q)]_{cos}$$
$$= [T_\pi(T_{(\pi/2+\delta)}(\cos + q)^2 G(\varepsilon\cos + \varepsilon q))]_{cos}$$
$$= -k(\mu, \delta, \varepsilon)$$

A similar calculation shows that h is odd. It follows that $k(\mu, \delta, 0) = 0 = h(\mu, \delta, 0)$.

We view system (B.12) as $G(\mu,\delta,\varepsilon)=0$ for $G:\mathbb{R}^3 \to \mathbb{R}^2$ satisfying $G(0,0,0) = 0$ and, as h,k are odd, $G(\mu,\delta,-\varepsilon) = G(\mu,\delta,\varepsilon)$. An easy computation shows that

$$\frac{\partial G}{\partial(\mu,\delta)}(0,0,0) = \begin{pmatrix} -1 & 0 \\ 0 & 1 \end{pmatrix}$$

The implicit function theorem implies that the equation $G = 0$ is solved by a C^1 function $(\mu,\delta) = (\mu(\varepsilon),\delta(\varepsilon))$ satisfying

$$\mu(-\varepsilon) = \mu(\varepsilon), \delta(-\varepsilon) = \delta(\varepsilon), \mu(0) = 0 = \delta(0)$$

Divide the second of equations (B.12) by ε^2 to get

$$\frac{\sin\delta(\varepsilon)}{\delta(\varepsilon)}\frac{\delta(\varepsilon)}{\varepsilon^2} = \frac{k(\mu(\varepsilon),\delta(\varepsilon),\varepsilon)}{\varepsilon}$$

Letting $\varepsilon \to 0$ and using that $k_\mu(0,0,0) = k_\delta(0,0,0) = 0$ we find that

$$\lim_{\varepsilon\to 0}\frac{\delta(\varepsilon)}{\varepsilon^2} = k_\varepsilon(0,0,0)$$

Similarly, dividing the first of equations (B.12) by ε^2, we get

$$-\frac{\mu}{\varepsilon^2}\frac{1}{1+\mu} + \frac{1-\cos(\delta)}{\delta}\frac{\delta}{\varepsilon^2} = \frac{h(\mu(\varepsilon),\delta(\varepsilon),\varepsilon)}{\varepsilon}$$

Taking the limit as before results in

$$\lim_{\varepsilon\to 0}\frac{\mu(\varepsilon)}{\varepsilon^2} = -h_\varepsilon(0,0,0)$$

The derivative $k_\varepsilon(0,0,0)$ can be computed from

$$k(0,0,\varepsilon) = [T_{\pi/2}(\cos+q(0,0,\varepsilon))^2 G(\varepsilon(\cos+q))]_{cos}$$

to be

$$k_\varepsilon(0,0,0) = 2G(0)[T_{\pi/2}(\cos q_\varepsilon(0,0,0))]_{cos} + G'(0)[\sin^3]_{cos}$$

$$= -2G(0)^2[\sin(z)(\frac{1}{2} + \frac{1}{10}\cos 2(z-\frac{\pi}{2}) - \frac{1}{5}\sin 2(z-\frac{\pi}{2}))]_{cos}$$

$$= -2G(0)^2[\frac{1}{2}\sin z - \frac{1}{10}\cos 2z\sin z + \frac{1}{5}\sin 2z\sin z]_{cos}$$

$$= -\frac{G(0)^2}{5}$$

Similarly,

$$h_\varepsilon(0,0,0) = -2G(0)^2[\frac{1}{2}\sin z - \frac{1}{10}\cos 2z \sin z + \frac{1}{5}\sin 2z \sin z]_{\sin}$$
$$+ G'(0)[(3/4)\sin(z) - (1/4)\sin(3z)]_{\sin}$$
$$= -2G(0)^2[\frac{11}{20}] + \frac{3G'(0)}{4}$$
$$= -\frac{11G(0)^2}{10} + \frac{3G'(0)}{4}$$

where we have used trigonometric identities such as

$$\cos(2z)\sin(z) = \frac{1}{2}[\sin(3z) - \sin(z)].$$

We summarize our computations as follows.

$$\delta = \delta(\varepsilon) = -\frac{G(0)^2}{5}\varepsilon^2 + o(\varepsilon^2) \tag{B.15}$$

$$\mu = \mu(\varepsilon) = \left(\frac{11G(0)^2}{10} - \frac{3G'(0)}{4}\right)\varepsilon^2 + o(\varepsilon^2) \tag{B.16}$$

Returning to the original parameters R, ω from (B.5), we have

$$\omega = 1 + \frac{2}{\pi}\delta$$
$$R = 1 + \frac{2}{\pi}\delta + \mu + \frac{2}{\pi}\delta\mu$$

which leads to

$$\omega = 1 - \frac{2G(0)^2}{5\pi}\varepsilon^2 + o(\varepsilon^2) \tag{B.17}$$

$$R = 1 + \left(G(0)^2\frac{11\pi - 4}{10\pi} - \frac{3G'(0)}{4}\right)\varepsilon^2 + o(\varepsilon^2) \tag{B.18}$$

Recall that
$$P = \varepsilon\cos(z) + \varepsilon q(\mu,\delta,\varepsilon)(z)$$

As q_μ and q_δ vanish at $(0,0,0)$ we have

$$q = q(\mu(\varepsilon),\delta(\varepsilon),\varepsilon) = \varepsilon q_\varepsilon(0,0,0) + o(\varepsilon)$$

and therefore, from (B.11), we have

$$P(z) = \varepsilon\cos(z) - \varepsilon^2 G(0)[\frac{1}{2} + \frac{1}{10}\cos(2z) - \frac{1}{5}\sin(2z)] + O(\varepsilon^3) \tag{B.19}$$

References

1. L. Ahlfors, *Complex Analysis*, McGraw-Hill, New York, 1966.
2. W. Aiello and H.I. Freedman, A time-delay model of single-species growth with stage structure, *Math. Biosci.* 101 (1990), 139–153.
3. D. J. Allwright, A global stability criterion for simple control loops, *J. Math. Biol.* 4 (1977), 363–373.
4. J. Arino, L. Wang, G. Wolkowicz, An alternative formulation for a delayed logistic equation, *J. Theor. Biology*, 241 (2006), 109–119.
5. J. Bélair, M. Mackey, J. Mahaffy, Age-structured and two-delay models for erythropoiesis, *Math. Biosci.* 128 (1995), 317–346.
6. R. Bellman and K. Cooke, *Differential-Difference Equations*, Academic Press, New York, 1963.
7. E. Beretta, F. Solimano, Y. Tang, Analysis of a chemostat model for bacteria and virulent bacteriophage, *Discrete Continu. Dynam. Syst.-B*, (2) (2002), 495–520.
8. F. Brauer, Absolute stability in delay equations, *J.Diff. Eqns.* 69 (1987) 185–191.
9. F. Brauer and C. Castillo-Chavez, *Mathematical Models in Population Biology and Epidemiology*, Springer, New York, 2001.
10. F. Brauer and J. Nohel, *The Qualitative Theory of Ordinary Differential Equations, An Introduction*, Dover, New York, 1989.
11. P. Brunovsky, A. Erdelyi, H.-O. Walther, On a model of the currency exchange rate-local stability and periodic solutions, *J. Dyn. and Diff. Eqns.* 16 (2004), 393–432.
12. S. Busenberg and K. Cooke, Periodic solutions of a periodic nonlinear delay differential equation, *SIAM J. Appl. Math.* 35 (1978), 704–721.
13. S. Busenberg and K.L. Cooke, The effect of integral conditions in certain equations modelling epidemics and population growth, *J. Math. Biol.* 10 (1980), 13–32.
14. S. Campbell, J. Belair, T. Ohira, J. Milton, Complex dynamics and multi-stability in a damped harmonic oscillator with delayed negative feedback, *Chaos* 5(4) (1995), 1–6.
15. S. Campbell and R. Jessop, Approximating the stability region for a differential equation with a distributed delay, *Math. Model. Nat. Phenom.* 4 (2009), 1–27.
16. W. Cheney, *Analysis for Applied Mathematics*, GTM 208, Springer , New York, 2001.
17. R. Churchill, *Complex Variables and Applications*, 2nd ed. McGraw-Hill, New York, 1960.
18. C. Colijn and M.C.Mackey, Bifurcation and bistability in a model of hematopoietic regulation, *SIAM J.Appl. Dyn. Systems* 6 (2007), 378–394.
19. K. Cooke, Stability analysis for a vector disease model, *Rocky Mount. J. Math.*, 9 (1979), 31–42.
20. K. Cooke and Z. Grossman, Discrete delay, distributed delay and stability switches, *J. Math. Anal. Appl.* 86 (1982), 592–627.
21. F. Crauste, Delay model of hematopoietic stem cell dynamics: Asymptotic stability and stability switch, *Math.Model.Nat. Phenom.* 4 (2009), 28–47.

22. R. Culshaw and S. Ruan, A delay-differential equation model of HIV infection of CD4+ T-cells, *Math. Biosci.* 165 (2000), 27–39.
23. R. Culshaw, S. Ruan, G. Webb, A mathematical model of cell-to-cell spread of HIV-1 that includes a time delay, *J. Math. Biol.* 46 (2003), 425–444.
24. J. Cushing, *Integrodifferential Equations and Delay Models in Population Dynamics*, Lect. Notes in Biomath.,#20, Springer-Verlag, New York, 1977.
25. A. De Gaetano and O Arino, Mathematical modeling of intra-venous glucose tolerance test, *J. Math. Biol.* 40 (2000), 136–168.
26. O. Diekmann, S.A. van Gils, S.M. Verduyn Lunel, H.-O. Walther, *Delay Equations: Functional-, Complex-, and Nonlinear Analysis*, Springer, Berlin 1995.
27. A. Dokoumetzidis, A. Iliadis, P. Macheras, Nonlinear dynamics in clinical Pharmacology: the paradigm of corticol secretion and suppression, *British J. Clinic. Pharmacol.* 54 (2002), 21–29.
28. S.F. Ellermeyer, Competition in the chemostat, asymptotic behavior of a model with delayed response in growth, *SIAM J. Appl. Math.* 54 (1994), 456–465.
29. S.F. Ellermeyer, J. Hendrix, N. Ghoochan, A theoretical and empirical investigation of delayed growth response in the continuous culture of bacteria, *J. Theor. Biol.*, 222 (2003) 485–494.
30. G. Enciso and E. Sontag, On the stability of a model of testosterone dynamics, *J. Math. Biol.*, 49 (6) (2004), 627–634.
31. K. Engelborghs, T.Luzyanina, G. Samaey, DDE-BIFTOOL v. 2.00: A MATLAB package for bifurcation analysis of delay differential equations, Technical Report TW-330, Department of Computer Science, K.U.Leuven, Leuven, Belgium, 2001.
32. C. Fall, E. Marland, J. Wagner, J. Tyson, *Computational Cell Biology*, Springer, New York, 2002.
33. T. Faria and L. Magalhaes, Normal forms for retarded functional differential equations with parameters and application to hopf bifurcation, *J.Diff.Eqns.* 122 (1995), 181–200.
34. G. Folland, *Real Analysis, Modern Techniques and Their Applications*, Wiley& Sons, New York, 1984.
35. T. Gedeon and G. Hines, Upper semicontinuity of Morse sets of a discretization of a delay-differential equation, *J. Diff. Eqns.* 151 (1999), 36–78.
36. L. Glass and M.C. Mackey,*From Clocks to Chaos*, Princeton University Press, Princeton NJ, 1988.
37. K. Gopalsamy, *Stability and Oscillations in Delay Differential Equations of Population Dynamics*, Kluwer, Boston, 1992.
38. S.A. Gourley, Y. Kuang, A stage structured predator-prey model and its dependence on maturation delay and death rate,*J. Math. Biol.* 49 (2004), 188–200.
39. I. Gyori and G. Ladas, *Oscillation Theory of Delay Differential Equations with Applications*, Oxford Science Publications, Clarendon Press, Oxford, 1991.
40. J.K. Hale, *Ordinary Differential Equations*, Krieger , Malabar FL, 1980.
41. J.K.Hale and S.M. Verduyn Lunel, *Introduction to Functional Differential Equations*, Springer, New York, 1993.
42. F. Hartung, T. Krisztin, H.-O. Walther, J. Wu, Functional differential equations with state-dependent delays: theory and applications, in *Handbook of Differential Equations* vol.3 (eds. A. Canada,P. Drabek,A. Fonda), Elsevier, Amsterdam, 2006.
43. B. Hassard, N. Kazarinoff, Y.-H.Wan, *Theory and Applications of Hopf Bifurcation*, Cambridge University Press ,Cambridge UK, 1981.
44. M.W. Hirsch and H.L. Smith, Monotone dynamical systems, in *Handbook of Differential Equations , Ordinary Differential Equations* (volume 2), eds. A.Canada, P.Drabek, A.Fonda, Elsevier, 239–357, Amsterdam, 2005.
45. Y. Hino, S. Murakami,T. Naito, *Functional Differential Equations with Infinite Delay*, Lect. Notes Math. 1473, Springer-Verlag, New York, 1991.
46. G. Hutchinson, Circular causal systems in ecology, *Ann. N.Y.Acad. Sci.* 50 (1948), 221–246.
47. J. Keener and J. Sneyd, *Mathematical Physiology II: Systems Physiology*, 2nd ed., Springer, New York, 2009.

48. Y. Kuang, *Delay Differential Equations with Applications in Population Dynamics*, Academic Press, New York, 1993.
49. M. Landry, S. Campbell, K. Morris, C. Aguilar, Dynamics of an inverted pendulum with delayed feedback control, *SIAM J. Appl. Dyn. Syst.* 4 (2005), 333–351.
50. B. Levin, F. Stewart, and L. Chao, Resource-limited growth, competition, and predation: a model, and experimental studies with bacteria and bacteriophage, *Amer. Naturalist* 111 (1977), 3–24.
51. R. Lenski and B. Levin, Constraints on the evolution of bacteria and virulent phage: a model, some experiments, and predictions for natural communities, *Amer. Naturalist* 125(4) 1985, 585–602.
52. J. Li and Y. Kuang, Analysis of a model of the glucose-insulin regulatory system with two delays, *SIAM J. Appl. Math.* (67) 2007, 757–776.
53. N. MacDonald, *Biological Delay Systems: linear stability theory*, Cambridge University Press, Cambridge UL, 1989.
54. P. Macheras, and A. Iliadis, *Modeling in Biopharmaceutics, Pharmacokinetics, and Pharmacodynamics Homogeneous and Heterogeneous Approaches*, Springer , New York, 2006.
55. J. Mallet-Paret, Morse decompositions for delay differential equations, *J.Diff. Eqns.* 72 (1988), 270–315.
56. J. Mallet-Paret and G. Sell, The Poincaré-Bendixson theorem for monotone cyclic feedback systems with delay, *J. Diff. Eqns.* 125 (1996), 441–489.
57. J..Mallet-Paret and H. Smith, The Poincaré-Bendixson theorem for monotone cyclic feedback systems, *J. Dynam. and Diff. Eqns.* 2 (1990), 367–421.
58. R. May, *Stability and Complexity in Model Ecosystems*, Princeton University Press, Princeton NJ, 1973.
59. W. Michiels and S.-I. Niculescu, *Stability and Stabilization of Time-Delay Systems, an Eigenvalue Approach*, SIAM, Philadelphia, 2007.
60. A. Mukhopadhyay, A. De Gaetano, O Arino, Modeling the intra-venous glucose tolerance test: a global study for a single-distributed delay model, *Discrete Cont. Dynam. Syst.-B* 4(2004), 407–417.
61. J.D. Murray, *Mathematical Biology*, Springer-Verlag, Berlin, 1989.
62. R. Nisbet and W. Gurney, *Modelling Fluctuating Populations*, Wiley-Interscience, New York 1982.
63. S. Ruan, Delay differential equations in single species dynamics, in *Delay Differential Equations with Applications*, ed. by O. Arino, M. Hbid and E. Ait Dads, NATO Science Series II: Mathematics, Physics and Chemistry, Vol. 205, 2006, Springer, Berlin, 477–517.
64. S. Ruan, On nonlinear dynamics of predator-prey models with discrete delay, *Math. Model. Nat. Phenom.* 4(2009), 140–188.
65. W. Rudin, *Principles of Mathematical Analysis*, McGraw-Hill, New York, 1976.
66. S. Saperstone, *Semidynamical Systems in Infinite Dimensional Spaces*, Applied Mathematical Sciences #37, Springer-Verlag, New York, 1981.
67. F. M. Scudo and J.R. Ziegler, *The Golden Age of Theoretical Ecology:1923–1940, A collection of the Works of V. Volterra, V.A. Kostitzin, A.J. Lotka, and A.N. Kolmogoroff*, Lect. Notes in Biomathematics 22, Springer-Verlag, Berlin, 1978.
68. H.L. Smith, Oscillations and multiple steady states in a cyclic gene model with repression, *J. Math. Biol.* 25(1987), 169–190.
69. H.L. Smith, Paul Waltman, *The Theory of the Chemostat*, Cambridge University Press, Cambridge UK, 1995.
70. H.L. Smith, Monotone semiflows generated by functional differential equations, *J. Diff. Eqns.*, 66 (1987), 420–442.
71. H.L. Smith, *Monotone Dynamical Systems: an introduction to the theory of competitive and cooperative systems*, Amer. Math. Soc. Surv. Monogra., 41, (1995).
72. H.L. Smith and H.R. Thieme, *Dynamical Systems and Population Persistence*, in preparation.
73. G. Stépán, *Retarded Dynamical Systems: Stability and Characteristic Functions*, Longman Scientific , 1989, Essex, England.

74. W. D. Stone, Weakly coupled systems of differential delay equations, Ph.D. thesis, University of Utah, 1983.

75. H. Thieme, *Mathematics in Population Biology*, Princeton University Press, Princeton, NJ, 2003.

76. P. van den Driessche, J. Wu, X. Zou, Stabilization role for inhibitory self-connections in a delayed neural network, *Physica D* 150 (2001), 84–90.

77. H.-O. Walther, On a model for soft landing with state dependent delay, *J. Dyn. Diff. Eqns.* 19 (2003), 593-622.

78. P. Waltman, *A Second Course in Elementary Differential Equations*, Dover, New York, 2004.

79. J. Wei and S. Ruan, Stability and bifurcation in a neural network model with two delays, *Physica D* 130 (1999), 255–272.

80. E. Wright, A nonlinear difference-differential equation, *J. Reine Angew. Math.* 494 (1955), 66–87.

81. E. Zeidler,*Nonlinear Functional Analysis and its Applications I:Fixed Point Theorems*, Springer-Verlag, New York, 1986.

Index